Safety Management Systems and their Origins

Safety Management Systems and their Origins: Insights from the Aviation Industry presents different perspectives on SMS to better decode what it means as a safety approach and what it implicitly conveys beyond safety.

The book uses the aviation industry as a basis for analyzing where the SMS stands in terms of safety enhancement. Through a socio-historical analysis of how SMSs emerged and spread across high-risk industries and countries, the book also explains the other stakes underpinning this new approach to safety management.

Features:

- Explores SMS as it is implemented in aviation based on examples from several countries and regions, namely the UK, USA, and Australia.
- Presents a socio-historical analysis of how SMSs emerged in high-risk industries.
- Provides insights to explain the existing limitations of SMS.
- Proposes new avenues to reach beyond the limitations of SMS.
- Discusses the COVID-19 pandemic within the framework of risk analysis.

The book is intended for safety professionals and regulators, as well as graduate students and researchers in safety science and engineering.

Safety Management Systems and their Origins

Insights from the Aviation Industry

Corinne Bieder

CRC Press is an imprint of the
Taylor & Francis Group, an **informa** business

Cover image: Shutterstock | Chinnapong

First edition published 2023
by CRC Press
6000 Broken Sound Parkway NW, Suite 300, Boca Raton, FL 33487–2742

and by CRC Press
4 Park Square, Milton Park, Abingdon, Oxon, OX14 4RN

CRC Press is an imprint of Taylor & Francis Group, LLC

ISBN: 978-1-032-30893-7 (hbk)
ISBN: 978-1-032-30894-4 (pbk)
ISBN: 978-1-003-30716-7 (ebk)

DOI: 10.1201/9781003307167

Typeset in Times New Roman
by Apex CoVantage, LLC

Contents

Figures

Tables

Abbreviations

ANSP	air navigation service provider
ATM	air traffic management
ATO	approved training organization
CAA	Civil Aviation Authority (generic name for national or regional civil aviation authorities but also name of the civil aviation authority in several countries, such as the UK and Singapore)
CASA	Civil Aviation Safety Authority (name of Australian CAA)
DGAC	Direction Générale de l'Aviation Civile (name of the French CAA)
EASA	European Aviation Safety Agency (name of the European CAA)
FAA	Federal Aviation Administration (name of the US CAA)
IATA	International Air Transport Association
ICAO	International Civil Aviation Organization
MRO	maintenance and repair overhaul
OEM	original equipment manufacturer
SMS	safety management system

Foreword

Corinne Bieder delivers here an astonishing text. We are faced with a rare dissection of a managerial tool, the safety management system (SMS), which is used in many high-risk industries and presented here in its aeronautical version. We are also invited to an intimate dive into the mysteries of the socio-technical and procedural world of aviation safety.

The strength of the subject lies in several points that deserve our attention. It is not often that we are offered a slow immersion into an ultra-technical world from the inside. Corinne Bieder is a specialist in aviation and an expert in aeronautical safety, aware of the most recent developments in human and organizational factors. This text comes from her doctoral thesis, which we had the honor of following closely. It took all the tenacity of its author to accept each new version to clarify and equip its reader with the essential knowledge allowing her to understand what is at stake in the deployment of an SMS. An undeniable pedagogy inhabits the whole text.

This book can be read as an ethnography of an object whose foundations draw on the frameworks of thought of engineering, risk and its calculation, and uncertainty and its acceptance. It also draws on the frameworks of thought of regulation science, which is confronted with the need to standardize organized safety practices. Describing, documenting, and understanding from the inside how such an object could find its way and respond to heterogeneous needs is the successful challenge of this book. But we have access to much more. We have access to the negotiations that prevailed at the birth of such a socio-technical equipment. In a way, its etiology is reported to us thanks to the collection of testimonies from frontline actors who participated in the establishment of the SMS in the aviation industry.

The way in which such a device functions is dissected with patience and method. Corinne Bieder explains what this philosophy, which is now at the heart of contemporary thinking on aviation safety, has brought and continues to bring. This managerial tool allows to link elements that are in different spaces and at different levels of aviation safety. The SMS articulates equipment, air operators, work organizations, and probabilities of occurrence. Along the way, the author also discusses the intrinsic limits of the object and questions its globalizing ambitions.

Finally, Corinne Bieder offers a third way of understanding. The SMS is not the whole of aviation safety, and it can neither be nor become so. There is an irreducible part of the social and technical worlds that manufacture air safety on a daily basis. Encapsulating the coordinates of this air safety in a systemic representation on paper could represent a form of utopia, an attempt to domesticate complex universes that have no room for error. But at the end of the road, the fragility of the enterprise is revealed. However, it helps to understand that the SMS is part of a line of attempts that all aim to reduce uncertainty.

The precision of the subject, the point of view of an insider, and a real ease in sharing sometimes unattractive knowledge on how this type of tool works make this book accessible and very useful. It also shows a willingness to inscribe it in a history of ideas housed in the heart of these complex socio-technical worlds. Finally, it represents a position taken by the author on the evolution of the thought of air safety that will allow the readers to orient themselves. One feels guided, initiated, and invited to dialogue on the same level, or almost. Aviation safety is a matter for experts, that's for sure. This book contributes to making it a public matter, a debate where everyone can feel concerned.

Mathilde Bourrier
Geneva, March 30, 2022

Foreword

This book discusses the safety management system (SMS), but it is above all a contribution to the sociological history of industrial safety, our collective desires, our contradictions, our successes, and the limits of these successes.

It unravels the mechanism that has prevailed in imposing a standard for the SMS in one of the safest industries in the world—the aviation industry—by reducing the SMS to its classical, reassuring, and stereotypical essence rather than battling to reexamine it or make it increasingly agile and adapted to the growing complexity of the real world.

The book also explains why the aviation industry only recently decided to embrace the principles of the SMS. The author details the expected benefit of a belated SMS adoption in an already ultra-safe industry, where the expected gain was not in the anticipated safety results (that have only marginally improved) but rather in demonstrating the standardized, visible, and reassuring nature of the approach, thus showing it to different stakeholders, professionals, users, and authorities. In other words, the SMS in the aviation industry was indispensable to the global development of the industry; however, its implementation was mainly motivated by a social and commercial agenda.

The editorial quality and the considerable expertise acquired by Dr. Corinne Bieder throughout her career make this book a reference work on the fundamental question of socio-technical games surrounding safety issues. The SMS provides a perfectly illustrated pretext to discuss the merits of two general visions of safety: *the first* is a prudent and conservative vision that aims to first and foremost reassure and use safety tools as ritualized objects, common to all, that serve as a transparent demonstration for stakeholders; *the second* is a more proactive vision, where the paper demonstration would be insufficient in the face of constant confrontation of reality, yielding an agile vision of the content, particularized to the context, at the risk of partially diminishing the reassuring nature of a common ritual.

To illustrate the richness of this debate and its complexity, which are so eloquently illustrated throughout the book, I will discuss four critical points of the debate within this short preface: (1) the safety discourse linked to the adoption of the SMS is rhetoric addressed to different audiences, with a primary focus for the less professional experts and ordinary citizens; (2) the SMS is limitless, which constitutes its fundamental trap; (3) the SMS is often excessive and misleading as its first priority is an easy and well-understood paper-and-pen demonstration to build client's and market's trust of the company or more generally the system; and (4) the SMS does not value uncertainty due to its alarming nature; nonetheless, this is a characteristic of the daily experience at work.

1. The safety discourse linked to the adoption of the SMS is a conservative rhetoric, a promise, addressed to different audiences, although primarily aimed at professionals with less expertise and ordinary citizens. As a result, the safety principles contain multiple readings and multiple intentions,

providing a collective framework of methods and actions for all types of professionals (particularly those with less expertise) and convincing and reassuring the public by using ambiguous terms to address these different targets.

Under these conditions, the author suggests that the SMS requirements are tempted to aim for the lowest common denominator accessible to all, with all the dissatisfaction associated with this idea.

It is important to recognize that the idea of the SMS was inherited by the historical tendency of supervisory authorities in the chemical, mining, and oil and gas sectors to delegate safety management to the manufacturers.

Since the '90s, there has been a significant expansion of air transport and a globalization and proliferation of air stakeholders of all sizes and maturity levels across the planet, which have compelled an already ultra-safe industry to adopt a safety solution emerging from several "less-safe" industries. The need to standardize, supervise, and reassure, more than to be innovative, is the first objective of adopting an SMS. As such, it calls into question the SMS's true aptitude for transforming the basis of risk management.

2. Safety obeys a logic of "always necessitating more," which, in its endless dimension, becomes notably unfair for professionals. As a result, any residual accident within an ultra-safe system is all the more unacceptable as it pinpoints a fault rather than an error as the tasks were formally planned and precautions were taken to avoid potential rifts. The issue lies with the exponential cost of the desirable actions required to eliminate the potential and exceptional residual accident, a cost that may jeopardize the economic balance of the system, resulting in a subject of arbitration for decision-makers.

   Professionals respond to this injustice and the unaffordable cost of ultimate safety by translating the issue in terms of legal liability and insurability. Industrial stakeholders seek an acceptable demonstration to external parties that is aligned with cultural expectations that, prior to the accident, they have behaved correctly and have acted to do the "right thing." This is the second main objective of the SMS, which represents a tool for "actor-games" that are socially devoted to reducing the legal vulnerability of the industry instead of dedicated to continuously evolving and improving.

3. Like any oath, the promise of added safety associated with the SMS is often excessive, with a tendency for a demagogic perimeter in the direction of common sense. The excellence of the current safety level in aviation inevitably complicates the calculated safety contribution in such a system.

   The principle of the SMS is to address the precursors that, when not respected, are thought to trigger accident paths; these precursors include reporting, proactive risk analyses, safe organizational approaches, audit tools, or specific defense barriers. This is the very heart of the SMS, which incidentally also constitutes its limit.

   This perimeter of precursors is often too limited to the elements associated with the aircraft use and techniques. We can foresee the content of the SMS evolving significantly if a more global systemic approach to the SMS were adopted; one that, for example, extends to the socio-political and cultural precursors, including the types of relationship between regulators and regulated, as well as the elements that are known to contribute to rare residual accidents. Such an approach would conceivably be more relevant for risk mitigation but, in the end, would impose a broader framework and would incur additional costs. It is part of the ongoing reality that arbitration has never been decided in this sense. As such, this simply adds to the limitations of the current SMS.

4. Last, safety does not value uncertainty due to its alarming nature. Nonetheless, this is a characteristic of daily work situations. There is no shortage of suitable academic frameworks to describe the safest organizations that deal with surprises and uncertainties (e.g., HROs, resilience engineering), and yet aviation does not implement such frameworks, partly because we lack the knowhow to combine them with more traditional frameworks and partly because the adoption of a more specialized and personalized solution for everyone would diminish the ubiquitous conviction of a universal and identical approach for everyone.

Dr. Corinne Bieder dives into convincing details, arguments, and subject matter of the challenges and limitations, including those previously detailed. The content is incredibly rich and composed of a combination of numerous technical and academic documents, including multiple stakeholder interviews and safety data from various countries and systems.

The author provides us with several original escape routes from the vicious circle of reducing the SMS design (so immutable so far) to the traditional approach to industrial safety. This is by far the most innovative part of the book—where each of these escape routes paved with obstacles is extremely well analyzed by the author.

In conclusion, this book is a mine of knowledge on safety issues that extend beyond the challenges of the SMS in aviation.

I must acknowledge I currently regularly employ the knowledge and lessons from this book while interacting in scientific discussions and conferences.

I greatly enjoyed reading this book and would undoubtedly encourage you to do the same.

Thank you, Corinne.

Happy reading!

Prof. René Amalberti, PhD
Director, FONCSI

1 Introduction

1.1 A PERSPECTIVE REFLECTING A DIVERSE BACKGROUND AND EXPERIENCE

Although fairly recent as a domain, with less than 50 years of existence, safety science embraces diverse topics and disciplines, giving rise to a variety of concepts, theories, and approaches. Therefore, stating that the following work is a safety science perspective on safety management or safety management system (SMS) would be an oversimplification. Even though the purpose is not to write a professional autobiography, some more detailed information about my background and experience will help the reader to understand the arguments that I make in this work.

Twenty-six years ago, I first trained as an engineer and then a risk manager. Feeling that the methods I had been taught were missing part of the picture, I rapidly turned to human factors, as they were called back then. I spent the first years of my career in an industrial research department working on the development of a human reliability analysis (HRA) method in the nuclear industry, together with colleagues from diverse backgrounds: engineering, ergonomics, and psychology. The challenges to understanding each other led me to undertake a degree in cognitive ergonomics, which put the "unique truth" and view I was familiar with from my engineering education into perspective.

Following this collective multidisciplinary work experience, I turned to consultancy (still with a flavor and part of the activity in applied research) in human and organizational factors and safety in a variety of high-risk industries (mainly in transportation, rail, road, air, but also touching upon energy or healthcare). Fifteen years ago, I joined the aviation industry with different posts related to human and organizational factors and safety, keeping external activities and relationships in the research world. After 11 years as an aviation industry insider, I turned to academic research in safety management in aviation.

This multifaceted background and experience have given me a longitudinal experience of safety management over 25 years, as well as an overview of the topic and the way in which it is tackled in diverse industries and diverse types of activities (industry, academia, consultancy, and regulatory bodies). Last, my background and regular exchanges with my growing network of colleagues from different disciplines in academia provided me with diverse perspectives on safety management and analysis of existing practices.

1.2 THE SMS IN AVIATION: STEP CHANGE OR DECOY?

As illustrated in Figure 1.1, safety in aviation has significantly improved since the middle of the last century, when commercial aviation started to dramatically expand, making this development possible.

DOI: 10.1201/9781003307167-1

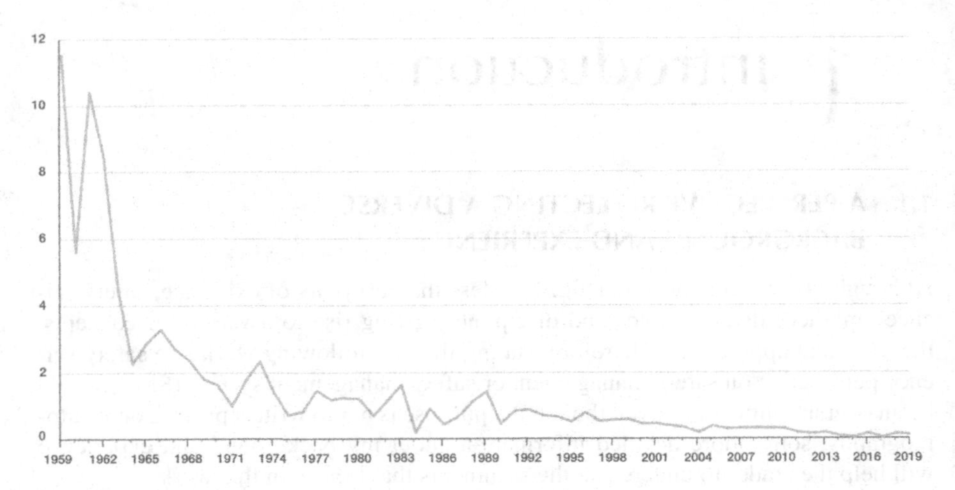

FIGURE 1.1 Yearly fatal accident rate per million flights.

Source: Airbus, 2020, p. 13.

Yet even though aviation is considered an ultra-safe system (Amalberti, 2001), safety remains a key concern. Exposure increases steadily and rapidly with traffic growth around the world, with air traffic doubling every 15 years, at least before the COVID-19 outbreak (see Figure 1.2).

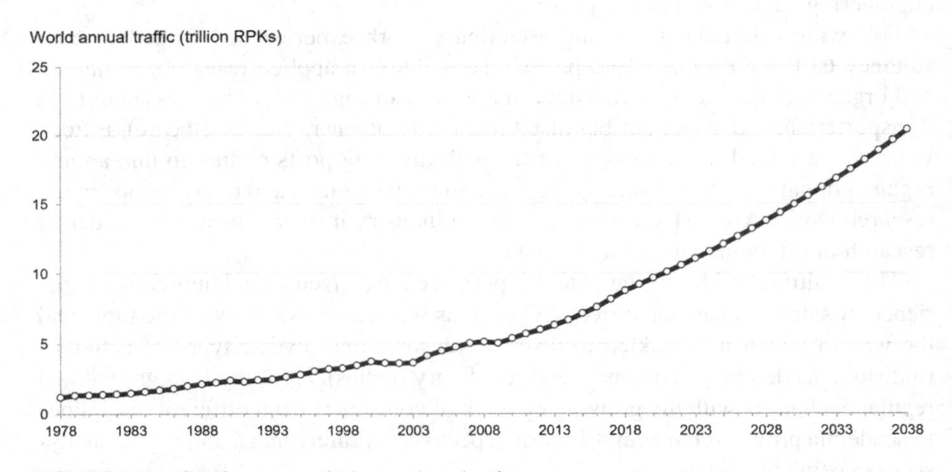

FIGURE 1.2 Traffic past evolution and projection.

Source: Airbus, 2019, pp. 10–11.

Moreover, safety issues may result in significant consequences, beyond accidents, for the organizations involved, as illustrated by the recent cases involving lithium batteries on the Boeing 787, leading to the grounding of the entire fleet for several months or the 737 MAX accidents follow-up leading to a halting of the aircraft production and operations, with a huge impact, and especially on the aircraft manufacturer.

In a context in which enhancing safety remains a priority, the introduction of the SMS as a game-changer to enhance safety in the mid-2000s conveys significant promises. Indeed, it is seen by many as the response to the new "total system" era, as stated by ICAO (see Figure 1.3).

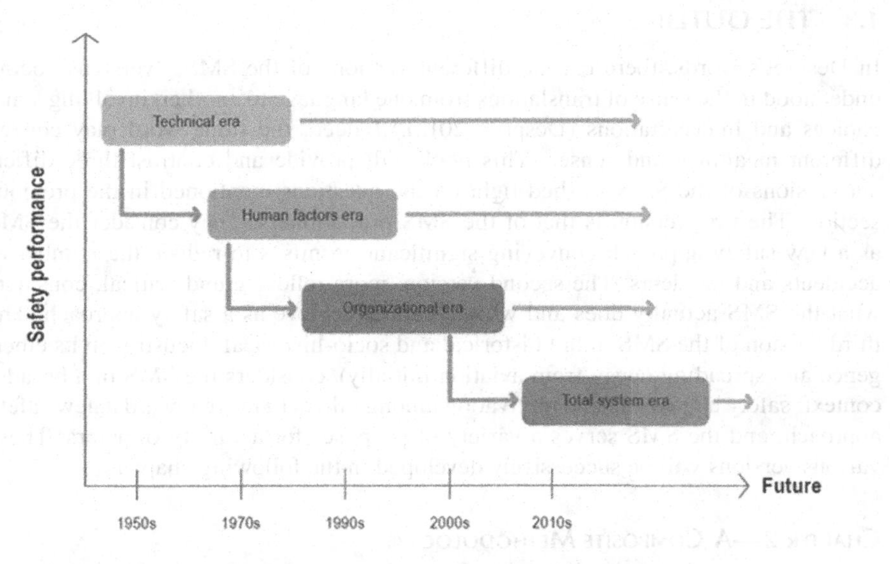

FIGURE 1.3 The evolution of safety as described by ICAO in the SMM.

Source: ICAO, 2018, pp. 2–2.[1]

This new era, according to ICAO, comes after previous ones where the focus was respectively on technology, human factors, and organizations, and the SMS appears as the new safety frontier. Following the preliminary report on the investigation of the Boeing 737 MAX accidents, more SMS for the aircraft manufacturer is advanced as a recommendation.[2] Interestingly, other high-risk industries, in particular chemistry and oil and gas, have developed and adopted SMSs much earlier (up to two decades before), at a time when the contribution of management to safety was starting to be theorized and disseminated as a way forward to enhance safety. Yet some major conceptual and theoretical developments in the organizational aspects of safety were yet to come. Conversely, aviation has come to the SMS after the organizational era and all the knowledge that was produced during the 1990s and early 2000s. Some confusion or shortcuts seem to have led almost naturally from the analysis of the role of organizational and managerial aspects in performing safely to SMSs as a response to take this new dimension into account.

Fifteen years later, doubts have been raised by practitioners and academics as to the effectiveness of the SMS in actually contributing to safety enhancement (Pelegrin, 2013; Dekker, 2014; Accou & Reniers, 2020). The objective of this work is to further investigate the case of the SMS in aviation. More specifically, the author will investigate how the SMS was widely adopted in aviation in the mid-2010s with the promise to be a step change in the management of safety and a way to reduce the occurrences of accidents. This involves exploring two key questions: How does the aviation SMS hold up to its safety enhancement promises? And how did the SMS land in aviation?

1.3 THE OUTLINE

In Despret's words, there can be different versions of the SMS, "versions" being understood in the sense of translations from one language to another, involving some choices and interpretations (Despret, 2012).[3] Indeed, the same word may convey different meanings and senses. This book will provide and contrast three different versions of the SMS to shed light on the questions mentioned in the previous section. The first version is that of the SMS practitioners: They consider the SMS as a new safety approach conveying significant promises to reduce the number of accidents and incidents. The second version, more reflexive and critical, considers what the SMS actually does and what it cannot achieve as a safety approach. The third version of the SMS, more historical and socio-historical, focusing on its emergence and spreading (away from aviation initially), considers the SMS in a broader context: safety appears as one motivation among others to move toward a new safety approach, and the SMS serves a variety of purposes for a variety of actors. These various versions will be successively developed in the following chapters.

Chapter 2—A Composite Methodology

Providing different versions of the SMS involves looking at it from different perspectives, through different lenses, with different frames in mind, but also using different methodological approaches. Chapter 2 provides an overview of the methods used to develop the different versions of the SMS. Although some methods served for all versions, like the literature review, even though the body of literature varied, some methods were more specific to some versions. For example, the first version describing the SMS as a safety approach full of promises relies primarily on document analysis. Likewise, the historical approach based on both the analysis of written sources and the interview of safety management old-timers, as well as the reflexive analysis close to auto-ethnography, mainly underlies the third version of the SMS, putting it back into a broader context where the actors' interests were diverse.

Chapter 3—Safety Actors' Version: The SMS as the New Safety Frontier

Chapter 3 provides the "official" version of the SMS. It describes the SMS as it is commonly presented and understood, namely, a new approach to managing safety in a proactive manner and considering the role of management in achieving safe performance. We extensively analyze the specific case of the SMS in aviation, starting from its promises and thoroughly reviewing what it looks like in practice. In order to perform an analysis that will be detailed enough to give a practical account of how the SMS translates in aviation organizations, the review is limited to the written traces of the SMS.

Chapter 4—What Does the SMS Actually Do, and Is It Up to Its Safety Promises?

The fourth chapter provides a more critical version of the SMS as a safety approach. Reaching beyond the promises, this version explores what the SMS in its aviation form actually does and, hence, what it may deliver—and, for either conceptual/theoretical, methodological, or practical reasons, what it cannot.

Chapter 5—Why Did the SMS Emerge and Spread?

Chapter 5 presents yet another version of the SMS, a socio-historical one, by considering the overall context, especially industrial, economic, and social, in which the SMS emerged essentially in the 1980s in the process industry. This version unravels the motivations of the main safety stakeholders to move toward a new safety approach, as well as the broader context elements that made an approach like the SMS an almost obvious minimal acceptable option to all parties. The SMS appears as a way to reconcile diverse motivations and diverse stakes in a specific industrial, regulatory, and social context, a kind of boundary object. The late adoption of the SMS in aviation is unpacked with this version in mind.

Chapter 6—Beyond the SMS: Toward More Contextualized Perspectives on Safety

Based on these various versions of the SMS, Chapter 6 suggests possible ways forward. Some of them are rather evolutionary and propose to reach beyond the main limitations identified from a safety viewpoint, especially by extending the approach to risk analysis. Others are more revolutionary and involve challenging the very consideration of safety as a stake one can "manage" in isolation from or even against others.

NOTES

1. Reproduced with the permission of ICAO.
2. Two Boeing 737 MAX aircraft crashed within five months. The first fatal accident, that of the Indonesian airline Lion Air flight 610, occurred on October 29, 2018, 13 minutes after takeoff from Jakarta. The second crash, killing as well all passengers and crews of the Ethiopian airlines flight 302, occurred on March 10, 2019. Both accidents occurred in similar conditions according to the preliminary report on the investigation of the Boeing 737 MAX accidents. They "shared a key contributing factor: a new software system called the Maneuvering Characteristics Augmentation System (MCAS), which Boeing developed to address stability issues in certain flight conditions induced by the plane's new, larger engines, and their relative placement on the 737 MAX aircraft compared to the engines' placement on the 737 NG" (House Committee on Transportation and Infrastructure, 2020, p. 1). The Boeing 737 MAX was certified by the US Federal Aviation Administration on March 8, 2017.

3. In the chapter "V for Versions," Despret distinguishes between two types of translations: prose, defined as "a translation where the primary value is accuracy and conformity with an original text" (Despret, 2012, p. 170), and versions, defined as "a translation that leads from another language back to its own, assumes, like any translation, some choices. In contrast to prose, however, these choices rest on the principle of a multiplicity of possible meanings, in the range of what is possible by 'homonymies': the same word can open up a number of meanings and different senses" (Despret, 2012, pp. 170–171).

REFERENCES

Accou, B., & Reniers, G. (2020). Introducing the extended safety fractal: Reusing the concept of safety management systems to organize resilient organizations. *International Journal of Environmental Research and Public Health*, *17*(15), 5478.

Airbus. (2019). *Global market forecast. Cities, airports & aircraft. 2019–2038*. Airbus.

Airbus. (2020). *A statistical analysis of commercial aviation accidents 1958–2020*. Airbus.

Amalberti, R. (2001). The paradoxes of almost totally safe transportation systems. *Safety Science*, *37*(2–3), 109–126.

Dekker, S. W. (2014). The bureaucratization of safety. *Safety Science*, *70*, 348–357.

Despret, V. (2012). V comme versions. V. Despret, Que diraient les animaux, si . . . on leur posait les bonnes questions, pp. 231–242.

House Committee on Transportation and Infrastructure. (2020, March). *The Boeing 737 MAX Aircraft*.

ICAO. (2018). *Safety Management Manual (SMM)* (4th ed.). Doc 9859. ICAO.

Pelegrin, C. (2013). The never-ending story of proceduralization in aviation. In: *Trapping safety into rules: How desirable or avoidable is proceduralization* (pp. 13–25). CRC Press.

2 A Composite Methodology

Exploring in-depth the multiple facets of the SMS in aviation, including the genesis of such a safety dispositive and its adoption in aviation, calls for addressing it from multiple angles, thus using complementary methods. Indeed, it involves not only an analysis of what the SMS is currently in aviation but also a historical approach to trace its origins and understand its emergence and dissemination. Studying the SMS in aviation today is a vast topic. Aviation involves a range of organizations such as airlines, airports, aircraft manufacturers, regulators, and air traffic management (ATM) organizations. Essentially, it is also an international activity, thereby embedded in national or regional economic, geographical, political, and social contexts. This complexity and variety create significant challenges for analysis. Besides, the SMS is not an isolated "object" that is easy to delineate. It also translates into practices that require extensive field work to be analyzed. Therefore, choices were made to scope out the study of the SMS in aviation today. As for the historical investigation, considering the relatively recent appearance of the SMS and the accessibility of actors and observers of its emergence and development, we adopted a socio-historical perspective. This chapter reports the methods that were relied on to study the SMS, as well as the reasons underlying these choices. The multiple angles and sources of data lead us to adopt a composite methodology.

2.1 DOCUMENT ANALYSIS: CONTRASTING SOURCES

Investigating the SMS in aviation can be done in several ways. Indeed, as further developed in Section 2.1, the SMS is a multifaceted "object" that articulates written documents and practices. Having access to SMS-related information (documentation, interviews, observations) requires us to develop a relationship close enough to build trust, especially when the request does not come from within the aviation organization itself. Furthermore, considering the diversity of aviation organizations, a sample would need to be big enough and international enough to be representative of the diversity of SMS embodiments. This involves significant challenges, including time and costs. These obstacles were circumnavigated by focusing the analysis on documents. According to Bowen (2009), the advantages of document analysis as a research method include its efficiency, cost-effectiveness, and coverage.

The author, throughout her career, came across a variety of written SMS sources, both public and confidential, and engaged in formal and informal conversations with SMS stakeholders from a variety of aviation organizations, including

DOI: 10.1201/9781003307167-2

authors of SMS documentation. Critically, the author had access to internal confidential SMS documents from various aviation organizations: the complete set of SMS documentation of two large aviation organizations, part of the SMS documentation of three more aviation organizations, and a benchmark between the SMSs of five aviation organizations, representing six different countries and three continents overall. The author also had the opportunity to follow the development process of SMS guidance material in one aviation regulatory organization before it became public. However, some of these sources are confidential and cannot easily be referred to.[1] Therefore, the focus of this research is on SMS documents produced by civil aviation authorities (CAAs). Beyond being public and easily accessible, guidance material and even more so resource kits also provide a good representation of the SMS as it is implemented in many aviation organizations, thereby responding to the availability and coverage advantages identified by Bowen (2009). First, they apply to the whole range of aviation organizations under the national or regional CAA. Second, many aviation organizations stick to the detailed guidance material rather than developing their own content (especially for resource issues as developed later). Therefore, this data selection choice allows us to overcome major confidentiality and sampling issues. Overall, the regulation, guidance material, resource kits, and web-based content on the SMS from six CAAs (Europe EASA and Eurocontrol, US FAA, UK CAA, French DGAC, and Australian CASA) and two international organizations (ICAO and IATA) were reviewed. However, the formal analysis underpinning the results presented in Section 3.2 mainly relies on the documents that provide the most detailed guidance on the SMS developed by three CAAs: the UK CAA, the US FAA, and the Australian CASA. This choice was driven by several aspects: (1) the language issue for the author to be able to read and review the documents, (2) the authority's understanding of the SMS (e.g., safety model, methodologies) reflected by the supporting documents, (3) the nature and level of guidance to support aviation organizations in their implementation of the SMS, and (4) the reach of the documents. The selected sources represent three different continents with a significant worldwide influence on safety management regulation and practices. All of these aspects are further developed in Section 3.2.3.

In Table 2.1, we present the main sources used as a basis for the formal document analysis.

2.2 LITERATURE REVIEW: SHADES OF GRAY

The first part of the work (presented in Chapter 3), focusing on the SMS as the new frontier to safety management, relied on a review of the scientific literature using the keywords "safety management system" or "SMS," "SMS" and "aviation," "safety management," and "risk management" on Scopus. Since aviation was among the late adopters of the SMS, the literature review starts long before the 21st century. The first publications on SMS appeared in the 1970s. However, to actually grasp what the SMS is in practice, especially in aviation, the review was extended to gray sources on safety management. According to Adams et al.

TABLE 2.1
Main Sources Used for the Formal Document Analysis

UK CAA	CAA. (2014). *Safety Management Systems (SMS) guidance for organisations*. Document CAP 795. CAA, London.	https://publicapps.caa.co.uk/modalapplication.aspx?appid=11&mode=detail&id=6616
	Web content accessible from the UK CAA web page dedicated to safety management systems, *Information for organisations regarding Safety Management Systems*.	www.caa.co.uk/Safety-initiatives-and-resources/Working-with-industry/Safety-management-systems/Safety-management-systems/
US FAA	Federal Aviation Authority (FAA). (2016). DRAFT AC 150/5200-37A Draft AC 150/5200-37A, *Safety Management Systems for Airports*.	www.faa.gov/documentlibrary/media/advisory_circular/draft-150-5200-37a.pdf
	Federal Aviation Authority (FAA). (2015). FAA AC_120-92B, *Safety Management Systems for Aviation Service Providers*.	www.faa.gov/documentLibrary/media/Advisory_Circular/AC_120-92B.pdf
Australian CASA	Civil Aviation Safety Authority (CASA). (2014a). *SMS for Aviation: A Practical Guide. Safety Management Systems Basics*. 2nd Edition. CASA. Canberra, Australia.	www.casa.gov.au/files/2014-sms-book1-safety-management-system-basicspdf
	Civil Aviation Safety Authority (CASA). (2014b). *SMS for Aviation: A Practical Guide. Safety Policy and Objectives*. 2nd Edition. CASA. Canberra, Australia.	www.casa.gov.au/files/2014-sms-book2-safety-policy-objectivespdf
	Civil Aviation Safety Authority (CASA). (2014c). *SMS for Aviation: A Practical Guide. Safety Risk Management*. 2nd Edition. CASA. Canberra, Australia.	www.casa.gov.au/files/2014-sms-book3-safety-risk-managementpdf
	Civil Aviation Safety Authority (CASA). (2014d). *SMS for Aviation: A Practical Guide. Safety Assurance*. 2nd Edition. CASA. Canberra, Australia.	www.casa.gov.au/files/2014-sms-book4-safety-assurancepdf
	Civil Aviation Safety Authority (CASA). (2014e). *SMS for Aviation: A Practical Guide. Safety Promotion*. 2nd Edition. CASA. Canberra, Australia.	www.casa.gov.au/files/2014-sms-book5-safety-promotionpdf

(2017), gray literature not only extends the range of evidence but also fills the gaps in the academic literature by contextualizing elements of research, thereby including "potentially relevant knowledge that is sometimes not reported adequately in academic articles" (op. cit. p. 438). Despite the fact that the process is not completely replicable, Adams et al. claim that gray sources still add value. The sources reviewed include both documents (combining public sources such as regulatory material and sources internal to aviation organizations, including SMS manuals and risk analyses) and discussions and meeting notes from exchanges with aviation safety practitioners, as well as with a mix of practitioners and academics, especially in the framework of the Future Sky Safety European project and other projects related to SMS and risk analysis conducted for aviation organizations.[2] The in-depth analysis of the content of actual SMS also benefited from my education and background in risk management.

2.3 A HISTORICAL APPROACH: WRITTEN SOURCES

The second and main part of the work, looking at the SMS from a broader perspective than that of safety, and especially focusing on how the SMS emerged as the new frontier to manage safety, requires a historical approach. Indeed, although the SMS was adopted in the early 21st century in aviation, the first discussions on the topic date from the 1970s. The systematic literature review mentioned earlier was then focused on the publications from 1970 to 2010. The references retrieved were analyzed not only in terms of abstract content (or detailed content when relevant) but also in terms of number and countries of origin to get insights into how ideas traveled.

We also explored archives such as conference programs when available, both recurring ones (e.g., Probabilistic Safety Assessment and Management conference, since the first conference in 1991) and ad hoc ones (e.g., World Bank workshop on safety control and risk management in 1988). More generally, archives from international organizations such as the World Bank, the FAA (Federal Aviation Administration), and the ICAO (International Civil Aviation Administration), available on the internet, were searched to trace references on safety management. The archive source turned out to be very limited in content. Indeed, there were no traces of the discussions and underlying rationales for adopting the SMS.

Interestingly, although primarily revolving around philatelic material, some high-level elements could be retrieved from the Postal History of ICAO web content, as illustrated in Figure 2.1.[3]

As for the World Bank, the most interesting material is a working paper covering the results of the workshop on organizational failures from the perspective of safety control and risk management held at the World Bank in October 18–20, 1998 (Rasmussen & Batstone, 1989). Although the detailed discussions are not transcribed, the document includes an overview of the papers that were presented, as well as the limitations and ways forward as they were identified and stated in the late 1980s.[4] Figure 2.2 provides an illustration of the way in which the problem was understood and stated through the list of presented papers.

"ICAO introduced the first version of the Global Aviation Safety Plan (GASP) in 1997 by formalizing a series of conclusions and recommendations developed during an informal meeting between the Air Navigation Commission (ANC) and industry. The plan was used to guide and prioritize the technical work programme of the Organization. It was updated regularly until 2005 to ensure its continuing relevance.

"In May 2005, another meeting between the ANC and industry identified a need for a broader plan that would provide a common frame of reference for all stakeholders. Such a plan would allow a more proactive approach to aviation safety and help coordinate and guide safety policies and initiatives worldwide to reduce the accident risk for commercial aviation. It was then decided that industry representatives (i.e. the Industry Safety Strategy Group (ISSG) whose members were Airbus, Boeing, Airports Council International, Civil Air navigation Services Organization, International Air Transport Association, International Federation of Air Line Pilots' Associations and Flight Safety Foundation), would work together with ICAO to develop a common approach for aviation safety. The Global Aviation Safety Roadmap that was developed by the ISSG provided the foundation upon which the expansion of the Global Aviation Safety Plan was based.

"While 2003 and 2004 were the safest years since the creation of ICAO in 1944, six major accidents in August and September 2005 claimed more lives than in all of 2004. From 20 to 22 March 2006, ICAO held in Montréal the Directors General of Civil Aviation Conference on a Global Strategy for Aviation Safety (DGCA/06), which welcomed the development of the Global Aviation Safety Roadmap and recommended that ICAO develop an integrated approach to safety initiatives based on the Global Aviation Safety Roadmap which would provide a global framework for the coordination of safety policies and initiatives."

International Civil Aviation Organization

DGCA CONFERENCE
2006
GLOBAL STRATEGY FOR AVIATION SAFETY

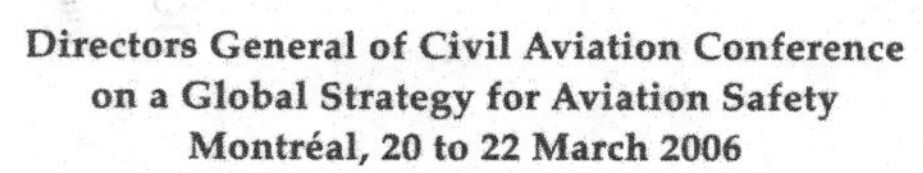

FIGURE 2.1 Extract from the Postal History of ICAO web content on safety management.

\- 1 -

1. LIST OF PAPERS PRESENTED

1.1 INTRODUCTION

1.1.1 **Rasmussen:** Workshop Outline Background Paper

1.1.2 **Rasmussen:** Some Concepts: A Note for Clarification

1.2 Industrial Safety Viewed as a Control Problem

1.2.1 **Sheridan:** Introduction: Industrial Safety Viewed as a Control Problem
1.2.2 **Rasmussen:** Industrial Safety Viewed as a Control Problem
1.2.3 **Volta:** Safety Control and New Paradigms in System Science
1.2.4 **LaPorte:** Discussion Paper
1.2.5 **Shikiar:** Discussion Points on Industrial Safety Viewed as a Control Problem
1.2.6 **Kugler:** Self-Organization and the Evolution of Instability: The Awakening of Sleeping Nonlinearities

1.3 Risk Management, Current Problems and Practices

1.3.1 **Reason:** Resident Pathogens and Risk Management
1.3.2 **Westrum:** Organizational and Inter-Organizational Thought
1.3.3 **Cramer:** A Programmatic Approach to Risk Management
1.3.4 **Ward:** Will the LOCA mind-Set be Overcome?
1.3.5 **Van Kuijen:** Prevention of Industrial Accidents in the Netherlands
1.3.6 **Cross:** Bank Failures
1.3.7 **Barreto-Vianna:** Risk Management, Current Problems and Practices in Brazil
1.3.8 **Thero:** Position Paper of the Quality Technology Company

1.4 Design of Reliable Organizations

1.4.1 **La Porte:** Safety Control and Risk Management:Views from a Social Scientist at the Operational Level
1.4.2 **Rochlin:** Technology, Hierarchy, and Organizational Self-Design. U.S. Naval Flight Operations as a Case Study
1.4.3 **Lanir:** Accidents and Catastrophes: The Safety Management of Contradictions
1.4.4 **Dynes:** Organizational Adaptations to Crisis: Mechanisms of Coordination and Structural Change (Including introductory note)
1.4.5 **Meshkati:** An Integrative Model for Designing Reliable Technological Organizations: The Role of Cultural Variables
1.4.6 **Fischhoff:** Simple Behavioral Principles in Complex System Design

\- 2 -

1.5 Organizational Decision Making

1.5.1 **Brehmer:** Changing Decisions about Safety in Organizations
1.5.2 **Kunreuther and Bowman:** Post Bhopal Behavior of a Chemical Company
1.5.3 **Ostberg:** On the Dilemma of High-level Decision-Makers in Their Handling of Risks in Political Contexts
1.5.4 **Baram:** The Influence of Law on the Industrial Risk Management Function
1.5.5 **Zimmerman:** The Government's Role as Stakeholder in Industrial Crisis

1.6 Adapting Risk Analysis to the Needs of Risk Management

1.6.1 **Brown and Reeves:** The Requirements in Risk Analysis for Risk Management
1.6.2 **Andow:** Discussion Paper
1.6.3 **Swain:** Some Major Problems in Human Reliability Analysis of Complex Systems
1.6.4 **Heising:** Discussion Statement, Two Pages
1.6.5 **Nertney:** Root Cause Analysis of Performance Indicators
Nertney: Adapting Risk Analysis to Needs of Risk Management, A Discussion Statement
1.6.6 **Wreathall:** Risk Assessment and its Use in Feedforward Control
1.6.7 **Embling:** The Risk Management System Approach
1.6.8 **Rowe:** Risk Analysis: A Tool for Policy Decisions
1.6.9 **Perdue:** A Decision-Analytic Perspective for Corporate Environmental Risk Management

1.7 Decision Support for Safety Control

1.7.1 **Rouse:** Designing for Human Decision Making in Large-Scale Systems
1.7.2 **Pew:** Human Factors Issues in Expert Systems
1.7.3 **Fussel:** PRISM: A Computer Program that Enhances Operational Safety
1.7.4 **Sheridan:** Trust and Industrial Safety
1.7.5 **Moray:** Can Decision Aids Help to Reduce Human Error?
1.7.6 **Woods:** Cognitive Error, Cognitive Simulation, and Risk Management

1.8 A Cross Disciplinary Discipline?

1.8.1 **Livingstone:** A Cross Disciplinary Discipline?

FIGURE 2.2 List of papers presented at the World Bank workshop on safety control and risk management in 1988.

Source: Rasmussen & Batstone, 1989, p. 1.

2.4 ORAL HISTORY: INTERVIEWING OLD-TIMERS OF SAFETY SCIENCE

Exploring how the SMS emerged and spread across the world and across high-risk industries (and eventually reached aviation) also requires completing the analysis of written sources with interviews. Indeed, as explained by the historian Descamps (2005), interviews give access to the implicit or invisible insiders' views on past situations and constitute strong bases for understanding how ideas traveled and developed. Beyond objective data, such as organizational charts, formal agreements or contracts between organizations, and the like, they provide insights into the actors and their backgrounds, interests, personal networks, and personal and professional paths that allow for making sense of the strategies implemented individually and collectively by the various actors. Interviews also allow for reaching beyond the official, declared objectives and strategies of organizations. Strategies are often reconstructed ex-post facto, although they emerge from multiple local actions that the confrontation of interviews allows for grasping and unveiling. Similarly, they allow for unveiling additional or hidden strategies behind the declared one.

Although written sources present a linear and rational story, interviews give access to "the complex interactions between the structure, the strategy, the actors and external environment permanently changing" (Chandler, 1989, cited by Descamps, 2005, pp. 578–579).

We conducted 18 interviews between July 2018 and September 2020. The interviewees were chosen for the personal part that they played in safety science or practices development before and/or when the SMS started to emerge either in their own domain or more globally. Overall, the sample of interviewees represents a range of safety management stakeholders and high-risk industries to provide a diversity of perspectives on the emergence of the SMS.[5] Interviewees were from regulatory bodies (five), high-risk industries (three), academia (four), and consulting companies on safety management (two), while four had hybrid profiles including two or more of these experiences either sequentially or simultaneously. Eight out of the sixteen were already involved in safety in the 1970s, eight started in the 1980s, and two in the 1990s. Considering the convergence of insights from several interviews to support our argument, the sample and the number of interviews seemed reasonable (Baker et al., 2012).

The interviews were open-ended, lasting between one and three hours. The framing of the scope was broad enough to gather as much information as possible. The purpose of the interviews was introduced as: an investigation into the origins of SMS, more specifically from a socio-historical perspective, that is focused on how ideas emerged, were discussed, and traveled. The scope and the objective of the interview were mentioned to the candidate interviewees well in advance (at least one week) at the time of the first contact. The first contact was made either by phone or by email.[6] When a convenient slot for the interview was agreed a confirmation email was sent, which reminded the interviewee of the purpose of the interview. The interviews were conducted in English for non-native French speakers and in French for others. The interviewees came from seven different countries and different places within the same countries. Whenever face-to-face interviews were not possible, interviews were conducted by phone. This turned out to be the simplest option since it was the most commonly shared and accessible among interviewees, including old-timers. Although one could think that it would undermine the quality of the interviews, some research outcomes prove it is not the case. After comparing face-to-face interviews and phone interviews, Sturges & Hanrahan (2004) concluded that there was no difference between the two modes in the data quality. Further, phone interviews may even be more suited than face-to face ones for qualitative research in some cases, especially when the topic is sensitive and the respondent prefers the anonymity of the telephone, when respondent groups are hard to reach, or when the costs are prohibitive. In our case, it allowed for gathering the views of people who would not have been reached otherwise, considering their location, thus enlarging the sample of interviewees but also diversifying the perspectives on the subject matter. Moreover, considering the historical nature of the topic of the interview, an immersion in the environment was not necessary. The interviews were not recorded, but extensive notes were made in real time. The notes were made to reflect as closely as possible the verbatim responses of the interviewees. The notes were then transcribed in a Word document that was sent to the interviewee for her/him to validate the accuracy of the interview transcription.

Table 2.2 provides an overview of the sample of interviewees, including insights on when the interviewee started to be involved in safety and from what region of the world they have been working.

Although some of the interviewees stated that they did not object to being named, some others explicitly asked for anonymity. In order to protect them all and handle all the interview material equally, all the quotes used later are anonymized. The interviewee profile is described as follows:

Professional activity: "academic," "industrial," "regulator," "safety management (SM) consultant," or "mixed experience" if the interviewee had different activities throughout his/her career

Type of industry: "chemistry," "nuclear," "aviation," or "diverse industries" if the interviewee worked in/for different industries throughout his/her career

2.5 QUALITATIVE CONTENT ANALYSIS: A MANUAL APPROACH FOR A LIMITED SAMPLE SIZE

A qualitative content analysis method was used to process the interview data since this historical investigation did not rely on a priori assumptions or theory (Descamps, 2005). Considering the limited size of the sample, the analysis was still manageable without software assistance. Although the interviews were a priori aiming at investigating how ideas traveled, all the interviewees referred as well to both the motivations of the various stakeholders to move toward a new approach to safety management and the overall context beyond safety that fostered the convergence

TABLE 2.2
Overview of the Performed Interviews and Insights on Interviewees' Profiles

	Interview date	In safety since	Reference including Region
1	Aug. 31, 2018	1990s	Regulator 1, aviation, Europe
2	Sep. 11–20, 2018	1970s	Academic 1, diverse industries, Europe
3	Sep. 25, 2018	1970s, maybe earlier	Regulator 2, aviation, America
4	Dec. 6, 2018	1990s	SM consultant 1, diverse industries, Europe
5	Dec. 21, 2018	1970s, maybe earlier	SM consultant 2, diverse industries, America
6	Dec. 28, 2018	1980s	Industrial 1, nuclear, Europe
7	Jan. 21, 2019	1980s	Mixed experience 1, diverse industries, Europe
8	Jan. 22, 2019	1970s	Mixed experience 2, diverse industries, Europe
9	Jan. 25, 2019	Early 1980s	Industrial 2, chemistry, Europe
10	Feb. 12, 2019	1990s	Industrial 3, aviation, Europe
11	Feb. 20, 2019	1980s	Regulator 3, aviation, Europe
12	Feb. 25, 2019	1980s	Regulator 4, aviation, Europe
13	June 14, 2019	1970s, maybe earlier	Academic 2, diverse industries, Europe
14	June 14, 2019	1970s, maybe earlier	Academic 3, diverse industries, Europe
15	July 23, 2019	Late 1960s	Mixed experience 3, diverse industries, Europe
16	June 18, 2020	1980s	Regulator 5, nuclear, Europe
17	June 29, 2020	1960s	Academic 4, diverse industries, America
18	Sep. 17, 2020	1980s	Mixed experience 4, oil and gas, Europe

toward the SMS. In the circulation and exchange of ideas, the following themes were identified inductively (Bengtsson, 2016):

Category	Subcategories
Within communities	Community of industrial field Community of users Scientific communities
Engines of transversality across communities	Individual personalities, trajectories, affinities Funding Accidents

The first category reflects that communities were a privileged circle within which ideas circulated.

The subcategories of communities identified may resemble the concept of "epistemic community" defined by Haas as "a network of professionals with recognized expertise and competence in a particular domain and an authoritative claim to policy-relevant knowledge within that domain or issue-area" (Haas, 1992, p. 3). However, the communities that were derived from the interview data did not meet the four criteria highlighted by Haas as characterizing epistemic communities, namely, (1) a shared set of normative and principled beliefs, (2) shared causal beliefs, (3) shared notions of validity, and (4) a common policy enterprise (ibid.). Some might have been epistemic communities in the making but not already there at the time the SMS emerged. Therefore, ad hoc categories were defined to reflect the data as far as possible. The community of the industrial field refers to people working in the same industry and sharing their knowledge and practices. The community of users refers to groups of people using the same method or tool and sharing their practices and feedback. As for the scientific communities, they build on Kuhn's definition whereby what makes the community is the sharing of similar paradigms (Kuhn, 1962).

The second category, engines of transversality across communities, reflects the mechanisms or conditions by which ideas could travel from one community to another. The subcategories of engines of transversality include conditions related to individuals but also to funding or to accidents.

The interview data was then processed again in the light of these (sub)categories to provide an understanding of how ideas related to safety management traveled.

2.6 A REFLEXIVE ANALYSIS OR AUTO-ETHNOGRAPHY

Part of the facts reported or underlying the observation and proposals put forward in this piece of work stem from my practitioner's experience and the variety of situations that I have had the opportunity to be involved in, observe, and/or discuss with a number of colleagues, clients, and peers as mentioned in Section 1.1. Beyond literature insights (both white and gray) and work performed in the framework of research projects, the following reflections are also based on thousands of meetings with a variety of practitioners and academics around safety management in many high-risk industries, thousands of hours of informal discussions with the same variety of people, and thousands of pages of notes, including verbatims.

Although not intended to be an auto-ethnographic research in the first place, this experience forms the background of the work presented in this document. Conversational interviews and observations were important inputs, even though I was a participant in the situations without any specific status identifiable from others more than a researcher becoming a participant observer as in auto-ethnography (Ellis et al., 2011). Whereas scientific observation and interview methods allow for grasping very rich insights into how organizations work, the longitudinal experience as a "full" insider within an organization provides a complementary perspective. The insider's experience involves not only a long exposure in time but also an intimate experience of the organization's structure and culture, as well as personal relationships, revealing some informal and unspoken aspects that may not be accessible to an external researcher. This positioning also offers a wealth of opportunities to hear off-the-record reactions and comments and the capability to interpret them in context and understand the various stakes. The same applies outside the organization, through relationships with other safety management actors. Furthermore, it allows for changing jobs, including managerial levels within the organization and viewing the organization from different perspectives, and hearing different discourses on safety management from within as well as outside of the organization.

My professional experience also took me through diverse industries and hazardous activities in industrial companies, small consulting companies, and academia, thereby diversifying my experiences in the safety management world. Just as gray literature allows for expanding insights and informs practices (Adams et al., 2017), an insider's industrial experience in several organizations, a consultant's experience for a range of organizations, and a researcher's experience in both the industry and academia allow heterogeneity to be captured and gives plural sources of practices, allowing to either corroborate or challenge some academic research outcomes, thus ultimately reopen scientific texts to discussion.

The extensive note-taking habit that I have developed over 26 years of work experience, grasping verbatims and personal analysis on the situation (e.g., reactions, tensions), provided me with significant material to document and analyze action on top of engaging in it (Anderson, 2006). As for my analysis of this insider's experience, it was further encouraged by my continuous interactions over the years with academics outside my organization that were interested in organizations and the impact of organizational practices (both internally and in their interactions with external parties) and culture on safety. These numerous exchanges engaged me in reflexive self-observation and regular analysis of my experience in the world of safety management.

NOTES

1. The SMS documentation includes, for example, a detailed analysis of the safety risks of an organization, as well as the thresholds determined by the organization to define the limits of risk acceptability.
2. This project, led by Eurocontrol from 2016 to 2019 and involving 35 participating organizations (aviation organizations, research organizations, consultants), received funding from the European Union's Horizon 2020 research and innovation program under grant agreement 640597.
3. See https://applications.icao.int/postalhistory/annex_19_safety_management.htm
4. Most of them are not accessible directly.
5. Some of the interviewees also shared some of their personal archives.
6. For two interviewees, the first contact was made face-to-face. Since the interviewer and the interviewees were physically located in the same place for a few days, it was decided that the interviews could be held face-to-face a day and a half later to let the interviewees think about the topic and prepare for the interview. The interviewees were reminded of purpose of the interview and the slots by email.

REFERENCES

Adams, R. J., Smart, P., & Huff, A. S. (2017). Shades of grey: Guidelines for working with the grey literature in systematic reviews for management and organizational studies. *International Journal of Management Reviews*, *19*(4), 432–454.

Anderson, L. (2006). Analytic autoethnography. *Journal of Contemporary Ethnography*, *35*(4), 373–395.

Baker, S. E., Edwards, R., & Doidge, M. (2012). *How many qualitative interviews is enough? Expert voices and early career reflections on sampling and cases in qualitative research.* Discussion paper, National Center for Research Methods.

Bengtsson, M. (2016). How to plan and perform a qualitative study using content analysis. *NursingPlus Open*, 2, 8–14. https://doi.org/10.1016/j.npls.2016.01.001

Bowen, G. A. (2009). Document analysis as a qualitative research method. *Qualitative Research Journal*, *9*(2), 27.

CAA. (2014). *Safety Management Systems (SMS) guidance for organisations.* Document CAP 795. CAA. https://publicapps.caa.co.uk/modalapplication.aspx?appid=11&mode=detail&id=6616

Chandler, A. D. (1989). *Stratégies et structures des entreprises*, MIT Press, 1962, Éditions d'organisation.

Civil Aviation Safety Authority (CASA). (2014a). SMS for aviation: A practical guide. In: *Safety management systems basics* (2nd ed.). CASA. www.casa.gov.au/files/2014-sms-book1-safety-management-system-basicspdf

Civil Aviation Safety Authority (CASA). (2014b). SMS for aviation: A practical guide. In: *Safety policy and objectives* (2nd ed.). CASA. www.casa.gov.au/files/2014-sms-book2-safety-policy-objectivespdf

Civil Aviation Safety Authority (CASA). (2014c). SMS for aviation: A practical guide. In: *Safety risk management* (2nd ed.). CASA. www.casa.gov.au/files/2014-sms-book3-safety-risk-managementpdf

Civil Aviation Safety Authority (CASA). (2014d). SMS for aviation: A practical guide. In: *Safety assurance* (2nd ed.). CASA. www.casa.gov.au/files/2014-sms-book4-safety-assurancepdf

Civil Aviation Safety Authority (CASA). (2014e). SMS for aviation: A practical guide. In: *Safety promotion* (2nd ed.). CASA. www.casa.gov.au/files/2014-sms-book5-safety-promotionpdf

Descamps, F. (2005). *L'historien, l'archiviste et le magnétophone: de la constitution de la source orale à son exploitation.* Comité pour l'Histoire économique et financière.

Ellis, C., Adams, T. E., & Bochner, A. P. (2011). Autoethnography: An overview. *Historical Social Research/Historische Sozialforschung*, pp. 273–290.

Federal Aviation Authority (FAA). (2015). *FAA AC_120-92B, safety management systems for aviation service providers.* www.faa.gov/documentLibrary/media/Advisory_Circular/AC_120-92B.pdf

Federal Aviation Authority (FAA). (2016). *DRAFT AC 150/5200-37A, safety management systems for airports.* www.faa.gov/documentlibrary/media/advisory_circular/draft-150-5200-37a.pdf

Haas, P. M. (1992). Introduction: Epistemic communities and international policy coordination. *International Organization*, *46*(1), 1–35.

Kuhn, T. S. (1962). *The structure of scientific revolutions.* University of Chicago Press.

Rasmussen, J., & Batstone, R. (Eds.). (1989). *Why do complex organizational systems fail? Summary proceedings of a cross disciplinary workshop on "safety control and risk management."* World Bank.

Sturges, J. E., & Hanrahan, K. J. (2004). Comparing telephone and face-to-face qualitative interviewing: A research note. *Qualitative Research*, *4*(1), 107–118.

3 Safety Actors' Version *The SMS as the New Safety Frontier*

Many different perspectives may hide behind the words "safety management system," even within the safety world. The SMS can be seen as "a move towards self-regulation" (Hale & Hovden, 1998); "a very practical concept, widely used in different industries" (Li & Guldenmund, p. 96); "a major research topic" (Hale et al., 1997, p. 121); or even "a new industry full of consultants" (Voss, 2012, p. 1). Nevertheless, all these perspectives seem to globally converge on what an SMS is intended to achieve and what it is overall. However, even if there is a relatively common wisdom on what an SMS is in principle, its translation into practice varies significantly from one actor to another and even more from one domain to another. This part starts with a general description of the SMS as it is commonly presented by safety actors. It then continues with a detailed description and analysis of the content of an SMS as it translates in practice in aviation. This chapter is intended to be as descriptive as possible and reports the content of an SMS as it is understood, required, and suggested to be implemented specifically by three CAAs. Contrasting these three sources allows us to illustrate the diversity of philosophies, methods, and levels of guidance underlying the SMS. However, the purpose of this chapter is just to give an account of what the SMS is in aviation and report various approaches to it rather than judging. A critical analysis will be presented in the following chapter.

3.1 WHAT IS AN SMS?—GLOBAL OVERVIEW

3.1.1 What Is an SMS Meant For?

The first published characterization of an SMS was provided by Kysor in 1973. He defined the basic objective of an SMS as to "provide the operator with a well-planned and organized system that will assist him, improve his safety level and attain his desired risk level" (Kysor, 1973, p. 99). Although other definitions were proposed by different scholars or regulators, as analyzed by Li and Guldenmund in their extensive literature review on the SMS (Li & Guldenmund, 2018), most definitions converge regarding the overall purpose of an SMS, namely, to achieve a "good" or desired safety performance. Despite very rare alternative considerations on the SMS, conventional wisdom supports that it is meant to enhance safety performance or maintain it at an acceptable level.[1]

DOI: 10.1201/9781003307167-3

3.1.2 What Is the Definition of an SMS?

According to Kysor in the early 1970s, an SMS is

> a planned, documented safety program that incorporates certain basic management concepts and activating elements into a well-organized safety system. The safety activity areas and supporting elements that comprise this system act and interact on one another to help achieve the desired safety level or risk level.
>
> **(Kysor, 1973, p. 98)**

The literature on the SMS is now abundant and combines both scientific publications and gray sources, such as regulatory material or industry documents. The detailed definition of an SMS varies slightly from one source to another, as highlighted by Li and Guldenmund (2018). However, based on a systematic review of the literature, the authors propose a general definition of an SMS, close to that of SMS actors: "An SMS is commonly defined as the management procedures, elements and activities that aim to improve the safety performance of and within an organisation" (Li & Guldenmund, 2018, p. 96). They also underline that beyond the words "safety," "management," and "system," other keywords characterize the broad meaning of an SMS whatever the context, namely, "activity, approach, control, operation, process and procedure" (Li & Guldenmund, 2018, p. 96). Along similar lines, Dekker, in his analysis of the foundations of safety science, defines an SMS as "a systematic approach to managing safety" that, among other things,

> defines how the organization is set up to manage risk; identifies operational risks and implements suitable controls; develops effective communications across all levels of the organization about these risks and their controls; implements a process to identify and correct nonconformities; supports a continuous improvement process.
>
> **(Dekker, 2019, p. 328)**

Eventually, despite the small variations in the definition of an SMS depending on the source and the time when it was proposed, both the purpose and the characterization of the SMS share a common understanding across domains among SMS stakeholders.

3.1.3 What Are the SMS Novelties and Promises?

Through the systematic approach to managing safety based on management procedures and controls, the SMS not only pretends to enhance safety but also to do it in a better way than before. The promises conveyed by this novel approach can be summarized as follows:[2]

- It helps get the management's attention to safety and better manage the resources allocated to safety.
- It reaches beyond a mere compliance-based approach to safety by being performance-based.

- It leaves each high-risk organization sufficient leeway to define the solutions best suited to its characteristics.
- It is a proactive approach to safety.

The idea of getting the management's attention to safety comes from various perspectives. In 1973, Kysor, coming from a management angle, makes a parallel between safety risks and financial risks that companies are willing to take financial risks to achieve business objectives, and suggests that it is "imperative that management approach its companies' safety and health activities in the same precise manner as it approaches its financial activities" (Kysor, 1973, p. 98), that is, through advanced financial management systems. Closer to safety as a core concern, other authors emphasize the role played by management in accidents, whether occupational or industrial ones (see, among others, Heinrich, 1931; Bird & Loftus, 1976; Reason, 1997; Vaughan, 1996). Although both ideas are, in reality, not fully similar, as discussed in the following chapter, they both come down to the involvement of management in safety.

As summarized by Swuste et al. (2018, p. 234),

> Accident reports, time and again, demonstrated that management failure was the fundamental cause of accidents, especially by major accidents, and thus adequate safety management systems were desperately required. Similarly, the scientific literature concluded that management decisions were often haphazard, leading to unsafe situations. This was found both for high-risk organisations and occupational safety. Safety simply did not get the attention it deserved from contemporary managers.

In this respect, the SMS builds on the concept of internal control that puts the emphasis on the manager's responsibility for safety and health (Hovden & Tinmannsvik, 1990). The authors claim as well that "management should pay as much attention to safety and health as to other control elements in the company, like productivity, economics, product quality, etc." (Hovden & Tinmannsvik, 1990, p. 23). Another aspect where the SMS builds on internal control is the leeway it pretends to leave to each organization "to adjust it to their own needs, routines, organisation and culture" (Hovden, 1998, p. 141). In other words, rather than relying on detailed prescriptions as to how to achieve a "good" safety level, the SMS remains at a meta-level of prescription by focusing on what an SMS should consist of. As stated by Kontogiannis et al. (2017, p. 128), "Safety management systems (SMS) are changing from a prescriptive style to a more 'self-regulatory' and 'performance oriented' model that is more proactive, participative and better integrated with business activities." The SMS can be considered performance-based in the sense that one of the key elements of an SMS is a risk control system (Hale, 2005) that allows focusing on performance by assessing the risk and comparing it against the objectives set by the organization. As developed in Section 3.2.3, the SMS includes a risk management approach and safety performance indicators that are monitored to drive safety enhancement measures. Before the advent of the SMS, regulatory requirements took the form of detailed specifications. Complying with these regulatory requirements was the main approach to ensure safety. Safety enhancement measures were based on accident or significant incident analysis. The SMS claims not to be reactive in the sense that it is also meant to rely on the analysis of less critical events to draw safety lessons (before any catastrophic or significant consequences occur) and derive safety enhancement measures. The focus on risk can also make the SMS a more proactive approach to safety by performing a priori risk analyses beyond the traditional feedback from experience and lessons learned approach.

3.1.4 What Does an SMS Consist Of?

In their literature review of the SMS, Li and Guldenmund reviewed 43 SMSs and compared their respective elements with Hale's generic SMS consisting of two main elements: the risk control system and the learning system (Hale, 2005). They concluded that SMSs varied for several reasons, especially (1) industries have specific safety management problems, (2) SMSs are more or less generic/specific, and (3) an element existing in two different SMSs could refer to different things (Li & Guldenmund, 2018). Thus, providing a generic characterization of what an SMS consists of turns out to be impossible if one wants to reach beyond the two main elements defined by Hale. Dekker proposes an illustration of what an SMS includes, namely, "organizational structures related to safety functions; people's responsibilities and accountabilities related to safety; processes for gathering, analyzing, and storing safety-related information; safety policies and procedures" (Dekker, 2019, p. 326). However, in order to perform an in-depth analysis of what an SMS consists of, one needs to focus on a specific industry, if not a specific organization.

3.2 THE SMS IN AVIATION: A CASE STUDY

A QUICK INTRODUCTION TO AVIATION

The perspective used to describe civil aviation is driven by the framework and the objective of the description. The purpose of this section is to provide a brief overview of the air transport system with an emphasis on its safety-related characteristics for readers who are not familiar with aviation.

Virtually everyone has his/her own idea of air transport, from a passenger's experience to multiple contrails visible in the sky. However, the reality of aviation operations (and what exists behind the scenes to keep these operations as safe as possible) is not necessarily well known and has evolved over time. Since the period of interest related to the adoption of the SMS by the aviation industry ranges from the late 1970s, when the first discussions on the SMS took place (in other domains), to a few years after the turn of the 21st century, when the SMS became a regulatory requirement in aviation, this section gives an account of the air transport system of that time.[3]

A Variety of Interconnected Stakeholders/Organizations

Performing a commercial flight involves a number of organizations interacting with one another in real time, but also a number of organizations (partly the same) interacting with one another sometimes years, if not decades, before the flight takes place. For instance, the aircraft manufacturer interacts with suppliers, engine manufacturers, certification authorities, and airlines at the time of the design of a new aircraft type that can then be operated for decades. To take one example, the Boeing 747 first entered into service in 1969 and is still operated today, more than 50 years later. Even though some novelties have been introduced since then, a great number of design choices were made and certified back in the 1960s. A brief description of the various stakeholders is provided in the Appendix. This includes, for each stakeholder, its role, both in real time and in the back office, as well as the nature of its interactions with others, with an emphasis on safety.

Figure 3.1 provides a slightly more complete picture also representing more remote stakeholders, not described in detail, but still involved in the safety of operations, especially suppliers and subcontractors that most of the aviation stakeholders mentioned previously call upon, or the state/government (e.g., transport ministry) or ICAO (International Civil Aviation Organization) that

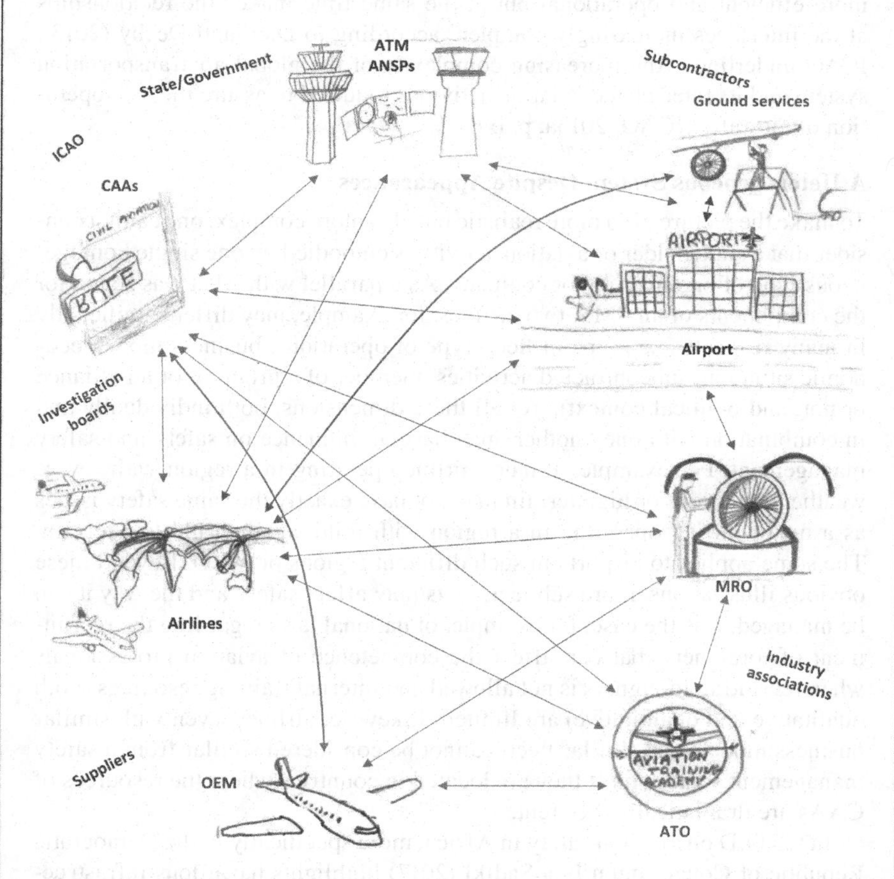

FIGURE 3.1 The main aviation stakeholders and their interactions, with an emphasis on operations and safety.

issues standards and recommended practices as to how to ensure safe operations that have a strong influencing power even though it has no legal value as such, or industry associations such as IATA, or investigation boards that may issue recommendations following accident investigations.

As such, aviation can be considered an interconnected infrastructure, or system-of-systems infrastructure. As highlighted by Gheorghe et al. (2006), in a system-of-systems infrastructure, (1) the overall structure escapes the control of any single actor, and (2) the different subsystems evolve autonomously. According to Harris and Stanton (2010), it would more simply be called a system of systems with multiple organizations, each having its own goals, interacting with one another to make commercial flights possible and safe. Maier (1998), cited by Harris and Stanton (ibid.), characterized a system of systems as possessing five basic traits: operational independence of elements, managerial independence of elements, evolutionary development, possessing emergent behavior, and having a geographical distribution of elements.

In addition, most of these organizations are fragmented, which makes them more efficient and operational but at the same time makes the relationships at the interfaces increasingly complex, according to Dien and Dechy (2013). ICAO underlines "the increasing complexity of the global air transportation system and its interrelated aviation activities required to assure the safe operation of aircraft" (ICAO, 2013a, p. ix).

A Heterogeneous System Despite Appearances

To make the picture even more realistic but also more complex, one cannot consider that a stakeholder or aviation activity is embodied by one single homogeneous type of organization or company. As a parallel with what was stated for the components of an SMS, two airlines, for example, may differ significantly in many respects (e.g., type of fleet, type of operations, business model, economic situation, subcontracted activities, member of a group or of an alliance or not, and political context). Yet all these dimensions, both individually and in combination with one another, may have an influence on safety and safety management. For example, a major airline operating in a region with severe weather conditions or high terrain may not have exactly the same safety issues as a major airline operating in a region with mild weather and low terrain. The same applies to airports in such different regions or MRO. Beyond these obvious illustrations, more subtle aspects may affect safety and the way it can be managed. It is the case, for example, of national laws regarding the recruitment of foreigners that can affect the competence of aviation professionals when recruiting foreigners is not allowed and internal training resources (both qualitative and quantitative) are limited. Likewise, airlines, even with similar business models and similar fleet, cannot be considered similar from a safety management standpoint if they are located in countries where the resources of CAAs are dramatically different.

In his PhD on aviation safety in Africa, more specifically in the Democratic Republic of Congo, Itabu Issa Sadiki (2017) highlights hazardous infrastructures such as runways, CAA's light controls, and complacency, leading to the delivery of undue operating certificates, approximate weather information due to lack of qualitative and quantitative resources of the weather services organization, as well as airlines operating aircraft in suspicious technical state flown by underqualified flight crews (op. cit., pp. 332–333). There is no such thing as an airline (or an airport, or a CAA) that would be representative of all the airlines even though they may share the same overall role/mission and interactions. Organizations have to be considered in their actual context rather than that of the overall industry (Grote, 2012).

Furthermore, companies having an international activity (e.g., international airlines, aircraft manufacturers) interact with a variety of governments and regulators, adding to the complexity of the socio-technical models presented previously. For example, an aircraft type needs to be certified by EASA and the FAA with two different regulatory frameworks (even though there are huge overlaps) to be authorized to fly in both Europe and North America.[4,5,6]

Last, aviation is not a stand-alone system operating in a void and should be considered in a broader context to understand its variety, as analyzed and illustrated by Merritt and Maurino (2004) (see Figure 3.2).

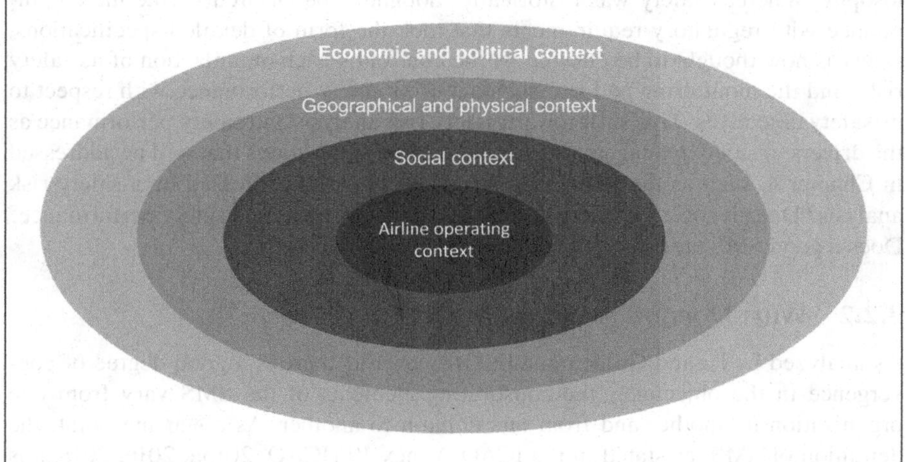

FIGURE 3.2 Aviation in context.

Source: Adapted from Merritt and Maurino, 2004, p. 151.

3.2.1 What Is the SMS Meant For?

As for the other high-risk domains that sometimes adopted the SMS long before aviation, the SMS is seen in aviation as an attempt to address safety in a more proactive way, to switch from a compliance to a performance-based approach to safety, and to turn safety into a business function to better get the attention of top management (Maurino, 2017; FAA website: www.faa.gov/about/initiatives/sms/explained; CASA, 2014a). Indeed, although historically safety has been ensured by compliance to rules, processes, and procedures, at least in transportation, the occurrences of accidents despite complying with rules called for another approach. Yet even if most aviation organizations did have safety management approaches in place, the generalization of the adoption of the SMS as such resulted from its becoming a regulatory requirement. One of the main shifts introduced by the SMS is a focus on safety risk management and safety performance as stated in the ICAO Annex 19, fully dedicated to safety management, issued in 2013:

> The Standards and Recommended Practices (SARPs) in this Annex are intended to assist States in managing aviation safety risks, with the objective of continuously reducing the number of aviation accidents and incidents. Given the increasing complexity of the global air transportation system and the interrelated nature of aviation activities required to assure the safe operation of aircraft, this Annex provides the means to support the continued evolution of a proactive strategy to improve safety performance.
>
> **(ICAO, 2013a, p. A7)**

As mentioned earlier, the novelty lies in the change in regulation and safety philosophy. Whereas safety was historically thought to be ensured by the mere compliance with regulatory requirements that took the form of detailed specifications, safety is now thought to be ensured by the control by each organization of its safety risks and the monitoring and management of its safety performance with respect to its safety objectives. This shift toward safety risk analysis and safety performance as the drivers of safety management raises a number of questions that will be addressed in Chapter 4, such as the following: What are the risks addressed by a safety risk analysis? Does it cover all safety issues? How do we apprehend safety performance? Does a good SMS ensure safe operations?

3.2.2 What Does the SMS Consist Of?

As analyzed by Li and Guldenmund (2018), even if there is a good degree of convergence in the objectives, the constituting elements of the SMS vary from one organization to another and from one domain to another. As a starting point, the definition of SMS, as stated in the ICAO Annex 19 (ICAO, 2013a, 2016, 2018a), is "a systematic approach to managing safety, including the necessary organizational structures, accountability, responsibilities, policies and procedures" (ICAO, 2018a, p. ix). However, the recommendations issued by ICAO and used as a basis by states to define their regulatory requirements go further in the description of what an SMS should consist of. The minimum requirements defined in ICAO Annex 19 (2013a, 2016, 2018a) include four pillars—(1) safety policy and objectives, (2) safety risk management, (3) safety assurance, and (4) safety promotion—and twelve elements that are generic to any aviation organization (ICAO, 2016, Appendix 2):

1. Safety policy and objectives
    - 1.1 Management commitment
    - 1.2 Safety accountability and responsibilities
    - 1.3 Appointment of key safety personnel

    1.4 Coordination of emergency response planning
    1.5 SMS documentation
2. Safety risk management
    2.1 Hazard identification
    2.2 Safety risk assessment and mitigation
3. Safety assurance
    3.1 Safety performance monitoring and measurement
    3.2 The management of change
    3.3 Continuous improvement of the SMS
4. Safety promotion
    4.1 Training and education
    4.2 Safety communication

3.2.3 What Does the SMS Look Like in Practice?

As mentioned earlier, the purpose of this section is to report how the SMS translates into practice in aviation. No judgment or critical analysis is provided at this stage. Yet some aspects may be more emphasized than others.

Heterogeneous Embodiments across Actors

As mentioned earlier, the SMS is not seen in the same way by all aviation stakeholders. Even if we focus on the written traces of the SMS (not to mention the organizational and practical aspects), the form taken by the SMS varies across actors. In a simplified way, whereas for CAAs, it is mainly a regulatory requirement and the associated guidance and supporting material, for aviation organizations, for aviation organizations/service providers it translates into a safety management manual (SMM) and the associated documentation (e.g., risk analyses, reported events). At a higher level, that of ICAO, the SMS is both an annex (Annex 19) presenting standards and recommended practices (SARPs) for safety management and a safety management manual providing states with guidance material and resource kits for them to implement effective state safety programs, including ensuring that aviation organizations implement the SMS as recommended in Annex 19.[7]

Generally, beyond the strict regulatory content, the guidance material provides additional insights on the "why," "what," and "how" questions. On the "how" to implement an SMS, the resource kits or supporting resources provide explicit examples of methods or contents or of "actual" SMS for fictitious organizations (see Figure 3.3).

Table 3.1 provides an illustration of the documentation structure associated with the SMS for (1) the ICAO at the international level, (2) several national or regional CAAs, and (3) aviation organizations.[8]

Overall, the SMS translates into the following:

- For regulatory bodies:
    - Legislation
    - Guidance material
    - Additional resources to support service providers more practically

TABLE 3.1
Illustration of the Documentation Associated with the SMS for the ICAO, National or Regional CAAs, and Aviation Organizations

Country	Organization	Document	Date of issuance	Status of document
International	ICAO	Annex 19[9]	2013; 2016	Standards and recommended practices[10]
		SMM (Doc 9859)	2006; 2013; 2018	Guidance material
Europe	European Commission[11]	European Aviation Safety Programme[12]	2015	Regulation
UK	CAA	CAP 1180 UK State Safety Program		Regulation
		CAP 795 Safety Management Systems (SMS) guidance for organisations	2014	Guidance material
		Bow-tie risk assessment models (including fully developed bow-tie templates)[13]		Additional supporting material available at the UK CAA website[14]
US	FAA	Safety Management Systems for Domestic, Flag, and Supplemental Operations Certificate Holders[15]	2015	Rule
		Safety Management System[16]	2013; 2016	National policy
Australia	CASA[17,18]	Parts of the CASR[19] Parts of the MoS		Legislation
		Advisory circulars Civil aviation advisory publications (CAAPs)[20] (e.g., CAAP SMS-01 v1.1 for Air Transport Operators)		Guidance material
		SMS: A Practical Guide, second edition	2014	Resource kit[21]
Every country around the world	Every aviation organization	Safety management manual		Internal documents explaining the organization's SMS (i.e., how the four pillars are implemented), shared with the oversight authority
		Documents associated with the SMS implementation, such as risk analyses and reported events		Internal documents, including confidential content

- For aviation organizations/service providers:
  - An SMS manual
  - A set of documents associated with the implementation of the SMS, especially risk analyses, reported events, and follow-up of safety performance indicators

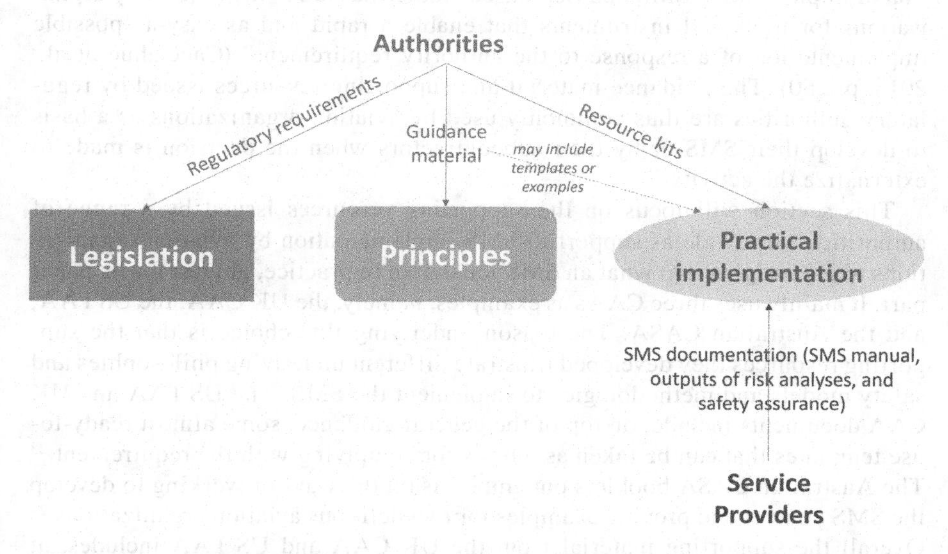

FIGURE 3.3 Illustration of the SMS documentation for authorities and service providers.

SMS as Understood and Presented in the Resource Kits: A Representative Perspective

Although the 12 elements of an SMS are common to all aviation organizations worldwide, their practical translation and embodiment are not. Performing a detailed analysis of SMSs actually developed by aviation organizations would pose huge challenges. Among the most critical ones are (1) the access to a sufficient sample of organizations worldwide and (2) the overcoming of confidentiality issues for each of these organizations. Indeed, as mentioned earlier, aviation organizations are diverse not only in terms of their activities but also in terms of their context. Furthermore, aviation organizations are generally not keen on revealing their risk analyses and evaluation and safety performance objectives or indicators. These would provide insights into their safety ambitions and their current strategies and results, including their weaknesses, in terms of safety. These challenges would exist even if the analysis were to remain at the level of the written traces of the SMS and disregard the actual practices beyond the paperwork. However, an interesting angle, allowing for circumnavigating this difficulty, is to focus on regulatory guidance material and even more on the resource kits produced by different CAAs. Even though this guidance material and other supporting resources do not have a regulatory status, they significantly influence (if not drive) the practices of most aviation organizations, especially those that do not have the necessary resources to perform their own SMS reflection and developments. As underlined by Cacciabue et al. (2015), performing a risk analysis and assessment requires a certain academic and practical training not commonly available within aviation organizations. Therefore, "the obligation of implementing SMS and risk-based studies has led to the request by organisations for tools and instruments that enable a rapid and as-easy-as possible implementation of a response to the authority requirement" (Cacciabue et al., 2015, p. 250). The guidance material and supporting resources issued by regulatory authorities are thus commonly used by aviation organizations as a basis to develop their SMS or by their subcontractors when the decision is made to externalize the activity.[22]

This section will focus on the supporting resources issued by a range of authorities worldwide as support to SMS implementation by aviation organizations to get insights into what an SMS looks like in practice, at least for its paper part. It mainly uses three CAAs as examples, namely, the UK CAA, the US FAA, and the Australian CASA. The reason underlying this choice is that the supporting resources they developed illustrate different underlying philosophies and safety models and methodologies to implement the SMS. The US FAA and UK CAA documents include, on top of the general guidance, some almost ready-to-use templates that can be taken as a basis for complying with the requirement.[23] The Australian CASA booklets put emphasis on the ways of working to develop the SMS content and provide examples for two fictitious aviation organizations.[24] Overall the supporting material from the UK CAA and US FAA includes, at one extreme, the prescriptive material and, at the other one, templates that can be used generically, and the Australian CASA provides prescriptive material (CAAP) and recommendations on how to proceed in practice to develop one's own content and examples. Whereas the first approach makes the copy-and-paste option quick and easy for organizations that do not invest much (whatever the reason) into the development of their SMS, the second one may induce more variety in the SMS content actually developed by aviation organizations (although they may also copy and paste existing material available from other sources including foreign CAAs). Besides the difference in the angle taken to support organizations, the content of the templates and examples developed by the various CAAs illustrate their safety philosophy, including their preferred methods and ways to implement them as discussed in detail in the following subsections.[25]

The supporting resources used as a basis for this analysis take different formats: documents on SMS in general like in the UK (e.g., CAP 795 SMS guidance or organizations; safety policy template), documents on SMSs dedicated to each aviation activity like in the US (e.g., FAA, DRAFT AC 150/5200-37A for airports; FAA, AC 120-92B for air carriers), or booklets for each of the four SMS pillars common to all aviation activities like in Australia (CASA, 2014a, 2014b, 2014c, 2014d, 2014e). The following subsections illustrate the form that each of the four pillars may take in practice and put into perspective the various ways of understanding and addressing how to implement this pillar when there are significant differences in the supporting resources of the three CAAs taken as examples.

Pillar 1: Safety Policy and Objectives: Ambition, Organization, Resources, and Commitment

Per ICAO Annex 19, the first pillar includes the five following elements:

1.1 Management commitment
1.2 Safety accountability and responsibilities
1.3 Appointment of key safety personnel
1.4 Coordination of emergency response planning[26]
1.5 SMS documentation

Management Commitment

This first element is addressed in similar ways by the US FAA and the UK CAA. They provide guidance that stays close to the original material developed by ICAO as to the elements a safety policy should include (see ICAO SMM, 2018b, p. 9–3). The recommended commitments include the improvement of the safety performance level in a continuous manner, the attention to the safety culture of the organization, the allocation of the necessary resources to ensure safety, the identification of safety as managers' primary responsibility, the compliance with regulatory requirements, the encouragement to report safety issues, and the dissemination of the safety policy at all levels. Furthermore, the FAA provides a "sample safety policy statement" in its guidance for SMS implementation for air carriers (see FAA AC 120-92B, Appendix) and the UK CAA a "template safety policy" (see UK CAA's webpage on SMS, section guidance and templates) that can be used as such.[27]

The Australian CASA Booklet 2 on Safety Policy and Objectives puts the emphasis on the establishment of expectations by the CEO, who is ultimately accountable for safety. "A safety policy outlines what your organization will do to manage safety. Your policy is a reminder of 'how we do business around here'" (CASA, 2014b, p. 1). Interestingly, "the way we do business around here" is commonly used as a definition of safety culture. Whereas both the US FAA and UK CAA guidance and examples refer to a reporting system or a reporting culture, the Australian CASA refers more generally to safety culture, both explicitly in the management commitment and responsibility checklist (see Figure 3.4) and through the examples provided that are a mix of safety and safety culture vision (see Figure 3.5 and Figure 3.6).[28]

In practice, while in some organizations, the safety policy is developed jointly by the CEO and a group of employees, in some others, the ready-to-use templates are reproduced as such with the company's name and logo added and signed by the CEO as any other document.

Management commitment and responsibility checklist

- There is commitment of the organisation's senior management to the development and ongoing improvement of the SMS
- This commitment should be demonstrated in a formal safety policy, which details:
  - The organisation's safety objectives
  - Management support of the SMS in providing the resources necessary for effective safety management
  - Who does what – a statement about responsibility and accountability for safety throughout the organisation
- There is evidence of decision making, actions and behaviours that reflect a more positive safety culture
- There is a defined disciplinary policy clearly identifying when punitive action would be considered (for example in cases of illegal activity, negligence or wilful misconduct)
- There is evidence that the organisation is applying its disciplinary policy

Conversely to the US FAA and UK CAA approach, CASA does not provide a safety policy template that could be used as such. They rather present a checklist not limited to the formal safety policy. Furthermore, they are neither prescriptive nor too specific as to what the safety policy should look like. It is up to the organization itself to reflect about it and enact its own

FIGURE 3.4 Extract of the CASA booklet on safety policy and objectives.

Source: CASA, 2014b, p. 1

Safety Accountability and Responsibilities

There is no significant difference in the way this element is addressed by the supporting material developed by the US FAA, UK CAA, or Australian CASA. All the documents remain very close to the ICAO SMM document.

The SMS leads to defining several lines of accountability:

- The CEO, typically the ultimate executive accountable for safety
- All the managers, accountable for the SMS activities within their area of responsibility
- Safety management personnel, including at least a safety manager[29]

Although the safety manager is responsible for developing the SMS manual and, more generally, implementing the SMS in all its dimensions and implications on behalf of the accountable executive, the safety manager is not the accountable manager.[30,31]

'We are all leaders in safety'

Safety leadership ultimately comes from the top, but everyone at Bush Aviation, regardless of position, can make a significant difference in reducing the number of near misses and accidents. Success in the future will depend on each one of us teaching, coaching and supporting others, so that no one is hurt, and no aircraft are damaged

Safety is the new economy

Safety is not just a priority, because priorities change; rather, it is our core and ever-present value. Our safety performance at Bush Aviation must continue to improve so we can lead our competition: in human performance, engagement and reduced worker turnover.

Think with both the heart and the head

Effective safety management is more than rules, training, safety meetings and a set of posters - those are just the mechanics. Everyone at Bush Aviation must believe that safety is important, make it automatic, and embrace it with all the energy, passion and persona! commitment it deserves.

The new team

Regardless of our role or professional background at Bush Aviation, we are all equal when it comes to safety responsibility - we are all in this together. I hope that you will join me in this exciting growth phase of Bush Aviation and make a valuable contribution to our safety system.

John Mathers

Chief pilot and owner
Bush Aviation and Training

FIGURE 3.5 Example of vision for Bush Aviation and Training.

Source: CASA, 2014b, p. 2

'At Outback Maintenance Services we have a proud safety record-not one of the aircraft we maintain has had an engineering accident, even a serious incident. But that doesn't mean we can sit back-far from it. We're only as good as our as last job. And you ail know what's happening with the mining in our area; that's where I think our future income is going to come from. I have talked to some of the people at Outback Exploration, and they have very demanding operational and safety standards.

'We have a vision and some procedures.

'I know safety is ultimately my responsibility, but to maintain and improve our safety performance, so that we can grow as the mining industry in our area grows, all of us have to make safety the basis of everything we do.

'As you know, Mick has put his hand up to be our safety officer, and we need to continue to work on our safety objectives, how it's all set up, and how we track how it's working.'

FIGURE 3.6 Example of vision for Outback Maintenance Services.

Source: CASA, 2014b, p. 3

The SMS also involves the development of a high-level safety committee (safety review board, or SRB) to oversee the SMS program, as well as for large organizations, one or more safety action groups, or SAGs (often per type of activity [e.g., operations, maintenance, ground services]) that implement the safety strategies coming from the safety committee. SAGs normally consist of managers and front-line personnel. As for the SRB, it involves the accountable executive, senior managers as well as the safety manager with an advisory role. "The SRB is strategic and deals with high-level issues related to safety policies, resource allocation and organizational performance" (ICAO, 2018b, p. 9–7).

The FAA guidance document for airports includes, for example, in a sample organizational chart for safety oversight, the accountable executive, SMS manager, and safety committee functions as illustrated in Figure 3.7.

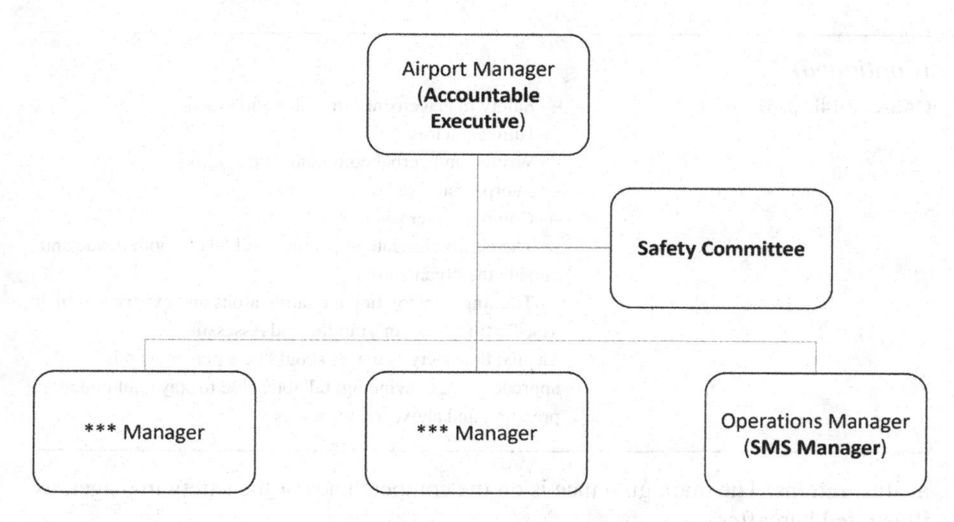

FIGURE 3.7 Main safety functions represented in a sample organizational chart for safety oversight as suggested in the FAA guidance document for airports.

Source: Adapted from FAA, 2016, p. 11

The organizational structure and the associated processes (at least the formal ones) are described in the SMS manual of an aviation organization. In practice, in large organizations combining several aviation activities (e.g., operations, maintenance, ground services), the safety organization comprises a safety manager at the corporate level and a set of safety managers, one for each activity, themselves relying on a network of professionals within the activity.

The formal structure and responsibilities described in aviation organizations' SMS manuals do not always translate into real practices. Some roles and responsibilities may, for example, be assigned to personnel already having other full-time duties (if not formally, at least in practice), especially in organizations that do not have a dedicated safety department with personnel exclusively dedicated to safety management.

Appointment of Key Safety Personnel

For this element, there is no significant difference either in the way in which it is addressed in the supporting material of the UK CAA or the Australian CASA, taken

UK CAA (2014, CAP 795, p. 7)	a) Broad operational knowledge and experience in the functions of the organisation and the supporting systems; b) Analytical and problem solving skills; c) Effective oral and written communication skills; d) An understanding of human and organisational factors; e) Detailed knowledge of safety management principles and practices.

(Continued)

(Continued)

CASA (2014b, p. 5)	— Safety management principles and practices — Human factors — Written and verbal communication skills — Interpersonal skills — Computer literacy — The ability to relate to people at all levels, both inside and outside the organization — Training—instructional qualifications and experience such as a Certificate IV in Training and Assessment Ideally, the safety manager should be a person who is approachable, convincing, reliable, able to stay cool under pressure, and above all, tenacious.

as illustrations. The main guidance is on the competencies of the safety manager, as illustrated hereafter.

In practice, other aspects come into play in the designation of the safety manager, especially in large organizations where career paths, grades, and other HR considerations may prevail. Figure 3.8 illustrates one of the examples of recruitment of key safety personnel (for a small organization) provided by CASA.

Bush Aviation Safety Officer | case study

The chief pilot drafts a position description for a part-time aviation safety officer, with the expectation that the role will grow with the business.

The chief pilot appoints one of his instructors, Patricia Chee, to the role and decides to send her on two training courses as a quick way to improve her safety management knowledge and skills, and build on her university study. By doing this, Mathers can utilise her skills and passion

1. Safety incident investigation training - two-day course
2. Human factors and error management - two-day course.

These courses cost $4,000, but the chief pilot feels this is a good investment in the future of his company.

The aviation safety officer arranges for one of CASA's aviation safety advisors (ASAs) to visit and provide further advice on SMS. Bush Aviation's safety officer receives some free CASA safety promotion products to further build the Bush Aviation SMS. The safety officer also attends free workshops and seminars on safety management hosted by CASA in the region.

Bush Aviation safety committee

Armed with the training Bush Aviation has provided, the safety officer establishes a safety committee, comprising the chief pilot/ CEO, one of the full-time pilots, the safety officer and one of the administration staff.

They also invite a representative from the local on-airport maintenance organisation, Outback Maintenance Services, to meetings.

FIGURE 3.8 Extract from CASA's guidance on the appointment of key safety personnel for a fictitious organization, Bush Aviation and Training.

Source: CASA, 2014b, p. 6

The US FAA does not expand much on the recruitment of the safety manager but rather elaborates on the designation of the accountable executive by providing a complex diagram, at least for air carriers (FAA, 2015, AC 120-92B, Appendix 4). In practice, the accountable executive is often the CEO.

SMS Documentation

Guidance as to how to document an SMS does not vary significantly from one CAA to another and is generally very close to the general ICAO recommendations. SMS documentation includes the following:

- An SMS manual provides a detailed description of the service provider's policies, processes, and procedures.
- Certain records substantiate the existence and ongoing operation of the SMS. Typically, these records include the outputs of safety risk management processes, the outputs of safety assurance processes, all safety-related training for each individual, and all safety communications.

Regarding the SMS manual itself, templates or even complete manuals for fictitious organizations can be found either on CAA websites or on safety organization web sites.

The UK CAA, for example, provides an SMS manual content page template built around the four pillars.[32]

The US FAA provides a sample SMS manual in the guidance document for airports that includes more than the structure since each section is partly developed but still needs to be completed with some statements specific to the organization (FAA, 2016, Appendix D).

A complete SMS manual for a fictitious organization, Xyz Air, was created by the EASA—EHSIT (European Helicopter Safety Implementation Team)—Specialist Team Ops. and SMS "to give to small/medium operators a reference in order to easily create their own SMS manual and organisation" (p. 2).[33]

In practice, the structure of SMS manuals is quite similar whatever the organization. It follows the four pillars and associated elements constitutive of the SMS as recommended by ICAO and reproduced by CAAs. As for the content, it is supposed to reflect the organizational structure and processes of the aviation organization. The SMS manual, as such, is largely procedural. As for the previous elements, some organizations reuse existing generic material like the sample SMS manuals mentioned previously, while others will define their own content.

What varies even more is the content of the records of the safety risk management, safety assurance, and safety promotion processes since the methods used to support the risk management process or the indicators to monitor the safety performance or the approach to safety training and communication may vary dramatically from one organization to another as developed in the following subsections on the second, third, and fourth SMS pillars.

Pillar 2: Safety Risk Management: The Cornerstone of the SMS

Per ICAO Annex 19, this pillar includes the following two elements:

2.1 Hazard identification
2.2 Safety risk assessment and mitigation

Although performing a proactive risk analysis on top of a retroactive one is a shared ambition of SMS, the guidance as to how to perform the safety risk analysis is not similar across CAAs. The common guidance is provided at a rather high level in the safety risk management process, as illustrated by the ICAO SMM (ICAO, 2013b) (see Figure 3.9).

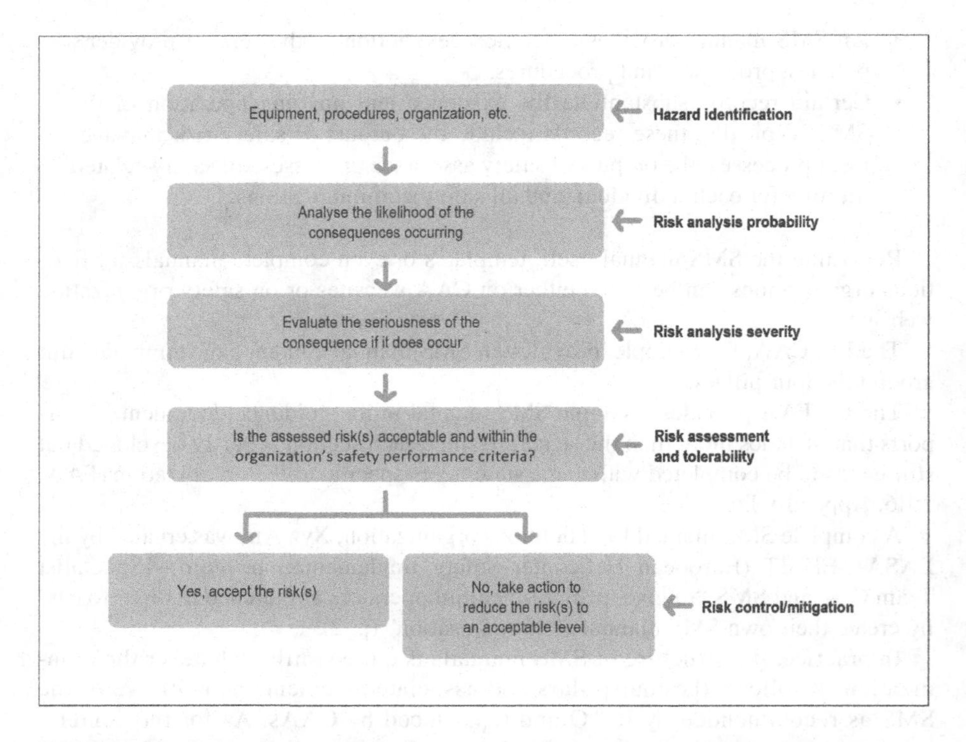

FIGURE 3.9 Guidance on hazard analysis and risk management process provided in the ICAO Safety Management Manual.

Source: ICAO, 2018b, pp. 9–11.[34]

At the level of detail of the two elements of this pillar, namely, hazard identification and safety risk assessment and mitigation, the existing supporting material goes into further detail, as illustrated hereafter.

Hazard Identification

The ICAO SMM (third edition) provides a generic list of hazards to consider (see Figure 3.10).

Beyond this generic list of hazards provided by ICAO, more detailed guidance and even examples are provided in the guidance material of national CAAs. The US FAA guidance document for airports mentions both the "possible sources of system failure," especially equipment, the operating environment, maintenance procedures and external services, and tools available for hazard identification, including "the Preliminary Hazard Analysis, Operational Safety Assessment, Comparative Safety Assessment, Fault-Hazard Analysis, What-If Analysis, Scenario Analysis, and the Fault Tree Analysis" (see FAA, 150/5200-37A,

"5.3.49 The following may be considered while engaged in the hazard identification process:

a) system description;
b) design factors, including equipment and task design;
c) human performance limitations (e.g. physiological, psychological and cognitive);
d) procedures and operating practices, including their documentation and checklists and their validation under actual operating conditions;
e) communication factors, including media, terminology and language;
f) **organizational factors, such as those related to the recruitment, training and retention of personnel, the compatibility of production and safety goals, the allocation of re-sources, operating pressures and the corporate safety culture**;
g) factors related to the operational environment <u>of the aviation system</u> (e.g. weather, ambient noise and vibration, temperature and lighting);
h) regulatory oversight factors, including the applicability and enforceability of regulations and the certification of equipment, personnel and procedures;
i) performance monitoring systems that can detect practical drift or operational deviations or a deterioration of product reliability;
j) human-machine interface factors;
k) factors related to the SSP/SMS interfaces with other organizations."

FIGURE 3.10 Hazards mentioned in ICAO documentation.

Source: ICAO, 2013b, pp. 5–17.[35]

p. 21). Beyond the general guidance, the FAA provides examples illustrating different types of hazards using different airport situations. One example (see FAA, 2016, p. 32) illustrates hazards rather related to the operational environment, namely, "incorrect aircraft parking at gate (caused by old markings)" and "misalignment of aircraft on ramp (caused by old markings)." Another example (see FAA, 2016, p. 34) illustrates hazards related to organizational aspects following a change of fixed base operator and, more specifically, a variety of "possible confusions" due to the "reduction in experience levels." Interestingly, the two examples also refer to two different ways of working to identify hazards—one rather individual, by the operations manager in the first example, and one more collective, with subject matter experts: "the airport manager decided to use a panel of subject matter experts that includes local pilots, the air carrier station manager, and members of the airport manager's staff" (see FAA, 2016, p. 34). In practice, both approaches are observed.

The UK CAA supporting material ranges from very high-level indications on the possible sources to identify hazards (CAP 795, pp. 12–13) to fully developed bow-tie templates that start with hazards, thereby providing a set of hazards.[36,37]

The Australian CASA puts emphasis on the ways of working within the organization to identify hazards. Rather than providing lists of either generic or detailed hazards or safety analysis methods, the guidance provides a combination of conceptual resources or general methods to support the organization's thinking about hazards applied to its own operations.

The guidance provided is described as follows (Figure 3.11):

There are several useful methods of identifying hazards:

» Brainstorming - small discussion groups meet to generate ideas in a non-judgmental way

» Formal review of standards, procedures and systems

» Staff surveys or questionnaires

» One person standing back from the operation and monitoring it critically and objectively

» Internal or external safety assessments

» Hazard reporting systems.

» Use of conceptual models such as:

- SHELL model
- Reason's accident causation model

Conversely to the ICAO, FAA or UK CAA guidance, CASA does not provide a list of more or less generic hazards, from which the aviation organization can derive its own ones. The guidance is far more upstream and provides possible approaches and ways of working for the organization to reflect by itself and identify its own hazards.

FIGURE 3.11 Guidance on hazard identification provided by CASA.

Source: CASA, 2014c, p. 1.

The CASA booklet also underlines the following:

> For larger organisations, setting up discussion groups with as many staff and line managers as possible is a good way of identifying hazards. The group discussions will also encourage staff to become more actively involved in establishing or improving your SMS.
> **(CASA, Booklet 3 on Safety Risk Management, p. 2)**

The CASA approach is here again to value the ways of developing the SMS as much as its final content, as part of the overall safety management approach.

Incidentally, there is no generic list of hazards or list of generic hazards provided in the CASA documentation on safety risk management. Another interesting feature of the CASA guidance is to call for "one person standing back from the operation and monitoring it critically and objectively." This reinforces the suggested non-naïve approach to controls and defenses in the risk analysis and management phase addressed hereafter.

In practice, the range of hazards identified by organizations varies depending on the sources they use, the ways of working, and supporting methods, if any.

Safety Risk Assessment and Mitigation

As for the hazard identification element, the nature and level of guidance to perform the risk analysis, assessment, and mitigation varies from one CAA to another.

Among the most detailed and easy-to-use guidance is that provided by the UK CAA, suggesting the use of bow-ties as a risk model for risk management.[38]

Illustrating this guidance material is particularly interesting since the bow-tie is very much used as a risk analysis and assessment approach supporting the SMS (Alizadeh & Moshashaei, 2015), including in aviation (Cacciabue et al., 2015). The UK CAA guidance material explicitly provides a complete set of bow-tie templates for the complete set of critical top events. It can be considered a relatively complete generic risk analysis. Although the risk analysis is supposed to be developed by the organization itself to reflect its operations and context, the bow-ties exemplified in the next paragraphs were developed

> in consultation with Subject Matter Experts (SMEs) from relevant sectors of the aviation industry in collaboration between the CAA and commercial operators. Extensive use was made of the workshop format with working groups of approximately six to eight participants guided by experienced bow-tie facilitators.[39]

The list of fully developed bow-tie models available on the UK CAA website includes the following topics further translated into more specific top events:[40]

- Airborne conflict, actually split into four bow-tie templates, each corresponding to more specific conditions (e.g., type of airspace)
- Controlled flight into terrain, split into three bow-tie templates
- Fire, split into three bow-tie templates, depending on the location of the fire
- Ground handling, split into three bow-tie templates related respectively to mass and balance, aircraft ground damage, or cold-weather-induced contamination of surfaces or engines
- Loss of control, split into four bow-tie templates (e.g., depending on the origin of the aircraft upset)
- Runway excursion, split into three bow-tie templates to cover longitudinal excursions respectively upon departure and landing, as well as lateral excursions
- Runway incursion, split into three bow-tie templates of what the runway incursion is interfering with

These bow-tie templates are developed around categories of accidents commonly shared within the industry—airborne conflict, controlled flight into terrain, fire, ground handling, loss of control, runway excursion, and runway incursion—to which are added three "umbrella" bow-tie templates to address the following: human performance, technical factors, and environmental factors. Each accident category is then split into several bow-ties corresponding to different top events that correspond to different ways in which the system can be destabilized and lead to this category of accident.[41] For example, a runway excursion (category of accident) can occur as the result of the inability to stop within distance when landing (bow-tie 2.1), or of a loss of directional control (bow-tie 2.2), or of aircraft acceleration or take-off differing from expectations (bow-tie 2.3).

Each bow-tie template is then fully developed from a CAA perspective—that is, integrating all the aviation actors, especially airlines, air navigation service providers (ANSPs), and airports. As an example, the bow-tie template for the top event "loss of directional control on the runway" associated with runway excursion suggests the hazards, preventative and recovery barriers, and the ultimate consequences (here only one) as follows (Figure 3.12):

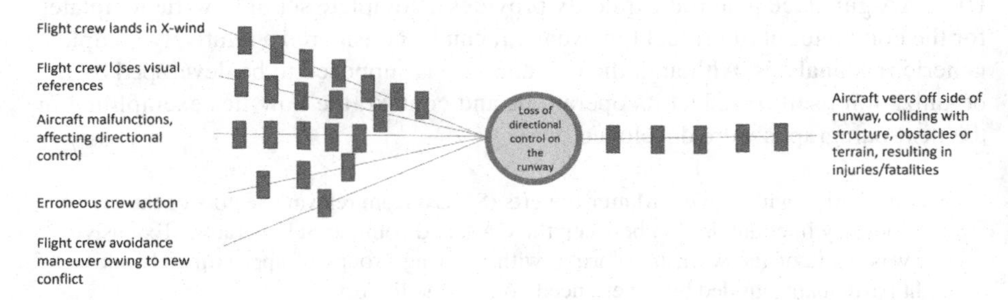

FIGURE 3.12 Bow-tie "loss of directional control on the runway" associated with runway excursion (each rectangle corresponds to a barrier).

Source: Adapted from the UK CAA website.[42]

Five hazards are considered:

1. Flight crew land in a crosswind (exacerbated by contaminated or slippery runways).
2. Flight crew lose adequate forward visual references for take-off/landing.
3. Aircraft malfunctions, affecting directional controllability during take-off/landing (e.g., thrust asymmetry, nose wheel steering).
4. Erroneous flight crew action on take-off/landing (e.g., rudder or accidental/asymmetric thrust).
5. Flight crew conduct avoidance maneuver owing to a conflict with another aircraft/vehicle on the runway.

The principle of bow-tie models is to make explicit the preventive barriers that are meant to stop each hazard from developing into the top event—that is, "loss of directional control on the runway" in the aforementioned example (Figure 3.13).

Similarly, further to the top event, the bow-tie model represents all the barriers meant to prevent the top event from degrading into the ultimate event—that is, "aircraft veers off side of runway colliding with structure, obstacles or terrain resulting in injuries/fatalities" (Figure 3.14).

In addition to these hazards and prevention and recovery controls, bow-ties also include escalation factors. These are defined as "a condition that leads to increased risk by defeating or reducing the effectiveness of controls (a control decay mechanism)."[43]

However, the UK CAA recommends not using overly-generic escalation factors to avoid "an explosion in the bow-tie diagram size without adding any useful information" (ibid.). In such a case, it is recommended to develop a separate dedicated bow-tie as in the case of fatigue.

The bow-ties exemplified previously, developed by the UK CAA with the support of industrial representatives, are meant to account for all aviation activities. In

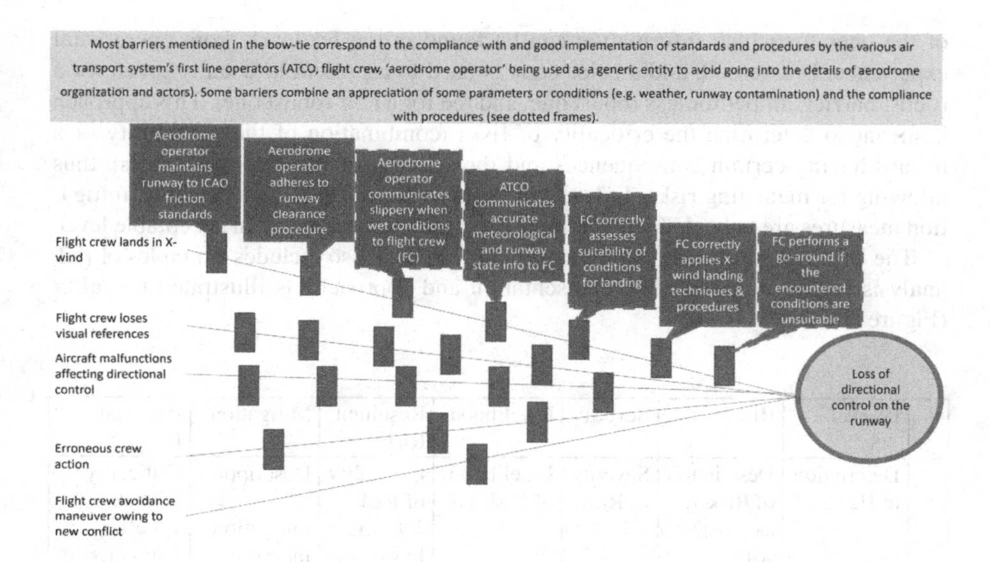

FIGURE 3.13 Detailed content of the preventative barriers for the first hazard of the bow-tie "loss of directional control on the runway" associated with runway excursion.

Source: Adapted from the UK CAA website.[44]

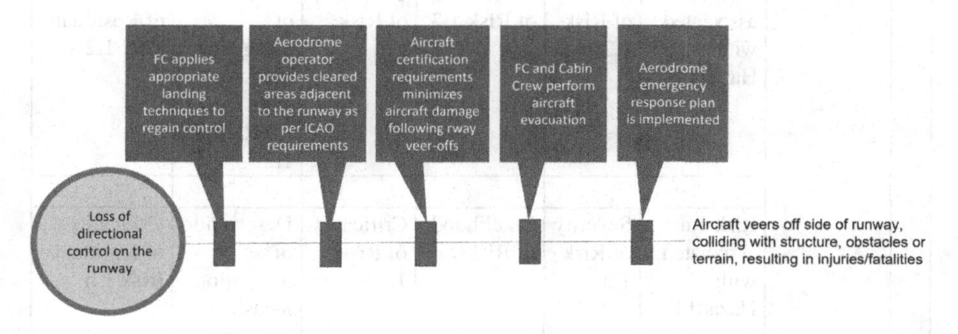

FIGURE 3.14 Detailed content of the recovery barriers of the bow-tie "loss of directional control on the runway" associated with runway excursion.

Source: Adapted from the UK CAA website.[45]

practice, even if the bow-ties are supposed to be developed by the organization itself to be adapted to its operations and context, some organizations reproduce available bow-ties with limited adaptation.

These bow-tie-modeling operations are then used to analyze feedback from operational experience, especially reported events. For example, an event reporting a loss of directional control on the runway, with ultimately no runway excursion, will be recorded in terms of the barriers that did hold and the ones that did not work. Ultimately, the robustness of the various barriers represented in bow-ties, or at least of the ones for which information can be found in the feedback from operational experience, is assessed, and a color code is used to represent it (usually green for a robust barrier, amber for less robust one, and red for a non-robust one). This approach is meant to determine the criticality of risks (combination of the probability of a hazard having certain consequences and the severity of these consequences), thus allowing for managing risks. Indeed, for the risks assessed as unacceptable, mitigation measures are to be defined to reduce them and bring them to an acceptable level.

The US FAA guidance documentation for airports also includes examples of risk analysis based on a different representation and approach, as illustrated hereafter (Figure 3.15).

#	Hazard	Risk	Severity	Likelihood	Resultant Risk	Mitigation	Residual Risk
1	Description of Hazard 1	Description of Risk 1.1 associated with Hazard 1	Severity of Risk 1.1	Likelihood of Risk 1.1	Criticality of Risk 1.1 (e.g. low-medium-high)	Description of mitigation measure envisaged for Risk 1.1	Criticality of Risk 1.1 after mitigation measure is implemented (i.e. residual Risk 1.1)
		Risk 1.2 associated with Hazard 1	Severity of Risk 1.2	Likelihood of Risk 1.2	Criticality of Risk 1.2	Description of mitigation measure envisaged for Risk 1.2	Criticality of Residual Risk 1.2
		…	…	…	…	…	…
		Risk 1.n associated with Hazard 1	Severity of Risk 1.n	Likelihood of Risk 1.n	Criticality of Risk 1.n	Description of mitigation measure envisaged for Risk 1.n	Criticality of Residual Risk 1.n
…	…	…	…	…	…	…	…
m	Description of Hazard m	Description of Risk m.1 associated with Hazard m	Severity of Risk m.1	Likelihood of Risk m.1	Criticality of Risk m.1	Description of mitigation measure envisaged for Risk m.1	Criticality of residual Risk m.1
		…	…	…	…	…	…

FIGURE 3.15 Risk analysis approach and format proposed in the FAA guidance material.

Source: Adapted from FAA, 2016, p. 34.

Rather than focusing on a specific top event and representing all the hazards likely to lead to this top event, the FAA approach starts from a hazard and identifies all the risks they can develop into, assesses each of them its criticality, and then identifies the mitigation that can take the risk back to an acceptable level.

The Australian CASA does not suggest or recommend any specific method for risk analysis in its supporting material. The only guidance is, once the hazards are ranked by criticality, to identify for each of the unacceptable ones, the defenses and controls in place to mitigate them. Yet CASA documentation recommends not to be naïve: "controls are never foolproof—for example, having well trained maintenance engineers does not ensure that aircraft components are always fitted correctly, and standard operating procedures for flight crew are only as effective as those who follow them" (CASA, Booklet 3 on Safety Risk Management, p. 2). This critical view on "work-as-done" versus "work-as-imagined" is reflected in the example provided in the safety risk management booklet as illustrated hereafter (Figure 3.16).

Sample hazard ID

Hazard ID	Shift handover & fatigue			
Mitigators	Effectiveness	Reason	Further controls/defences required	Responsibility
Shift handover procedures	No	In a manual in Peter's office – nobody reads them	Half-hour overlap between shifts to allow for proper briefing, and for log to be fully completed	Cheryl Lawson
Shift handover log	No	Not in central enough place – goes missing	To be transferred to hangar PC, and completed online	Cheryl Lawson
Regular staff safety meetings	No	Not held consistently enough	Schedule regular fortnightly toolbox meetings.	Mick xxx (safety officer)
Rostering	No	Not enough staff to cover the required shifts	With planned growth, take on new staff	Peter Lawson
Recording	No	Ad hoc system – is only done sometimes	Hazard & risk register on hangar PC. Everyone gives and receives feedback	Mick xxx & Cheryl Lawson

FIGURE 3.16 Screenshot from the CASA guidance material booklet on the shift handover and fatigue hazard.

Source: CASA, 2014c, p. 21.

This example illustrates the critical view on procedures and processes. Indeed, even though shift handover procedures or regular safety meetings are considered mitigators of the shift handover and fatigue hazard exemplified in Figure 3.16, neither of them is considered efficient in the current situation, which leads the organization (Outback Maintenance Services) to consider other barriers such as a half-hour overlap between shifts in the first case or regular fortnightly toolbox meetings in the second case. In other words, proposed barriers reach beyond the sole reinforcement or enforcement of compliance with a procedure or proceed to envisage more realistic risk reduction measures.

In practice, the safety risk management pillar implementation varies significantly from one organization to another. One reason for this lies in the investment the organization decides to put into the SMS, as well as the associated ways of working. Especially choosing to develop one's own risk analysis or performing a cut-and-paste of the bow-ties available with minimal adaptation will lead to very different outputs. Likewise, a safety manager doing a safety risk analysis on his/her own may not lead to the same content as when involving other personnel. Indeed, a risk analysis is as rich as the knowledge, experience and imagination of the ones who perform it. Another reason lies in the method(s) used to support the risk analysis. As illustrated before, a bow-tie does not look like a risk analysis as that exemplified by the FAA or CASA. Beyond the format, the nature of the content is different. As mentioned earlier, bow-tie is one of the most commonly used methods to support risk management in aviation.[46]

Pillar 3: Safety Assurance: Where Safety Models Underlying the Understanding of Safety Performance Unveil

Per ICAO Annex 19, this pillar includes the two following elements:

3.1 Safety performance monitoring and measurement
3.2 The management of change
3.3 Continuous improvement of the SMS

The safety assurance pillar aims at monitoring that the SMS works well. As defined by ICAO, it

> consists of processes and activities undertaken by the service provider to determine whether the SMS is operating according to expectations and requirements. The service provider continually monitors its internal processes as well as its operating environment to detect changes or deviations that may introduce emerging safety risks or the degradation of existing risk controls. Such changes or deviations may then be addressed together with the safety risk management process.
>
> **(ICAO, 2013b)**

The implementation of this pillar in practice varies dramatically from one organization to another.[47]

Safety Performance Monitoring and Measurement

Safety performance monitoring and measurement aims to measure the effectiveness of safety risk controls according to the ICAO SMM. It represents the feedback loop between the safety objectives and the current level of the organization with respect to these objectives. As stated by CASA in its booklet on safety assurance (CASA, 2014c, p. 1), "Safety assurance is the way you demonstrate your SMS works." As such, once an organization has defined its safety objectives, safety performance monitoring mainly relies on the definition of relevant indicators (i.e., those that are meant to reflect the safety performance of the organization) and the monitoring of whether these indicators are meeting their targeted level.

Although the guidelines for establishing safety objectives is limited (objectives are supposed to be specific to the organization), there is more material supporting the definition of safety performance indicators (SPIs) (that can, in turn, be used by organizations to define their objectives by setting a targeted level for some of the indicators provided as examples). Among the commonly used SPIs are, for example, the number of accidents, the number of incidents, the number of safety reviews, and the number of safety trainings delivered.[48]

The third version of the ICAO SMM provided examples of high-consequence indicators and lower-consequence indicators for each of the aviation activities that require to have an SMS.[49] Although some SPIs are specific to some activities, the general philosophy of SPIs is similar across all activities.[50]

Figure 3.17 illustrates the SPIs proposed by ICAO for air operators.

SMS safety performance indicators (individual service provider)					
High-consequence indicators (occurrence/outcome-based)			*Lower-consequence indicators (event/activity-based)*		
Safety performance indicator	*Alert level criteria*	*Target level criteria*	*Safety performance indicator*	*Alert level criteria*	*Target level criteria*
Air operator individual fleet monthly serious incident rate (e.g. per 1 000 FH)	Average + 1/2/3 SD (annual or 2 yearly reset)	__% (e.g. 5%) improvement between each annual mean rate	Operator combined fleet monthly incident rate (e.g. per 1 000 FH)	Average + 1/2/3 SD (annual or 2 yearly reset)	__% (e.g. 5%) improvement between each annual mean rate
Air operator combined fleet monthly serious incident rate (e.g. per 1 000 FH)	Average + 1/2/3 SD (annual or 2 yearly reset)	__% (e.g. 5%) improvement between each annual mean rate	Operator internal QMS/ SMS annual audit LEI % or findings rate (findings per audit)	Consideration	Consideration
Air operator engine IFSD incident rate (e.g. per 1 000 FH)	Average + 1/2/3 SD (annual or 2 yearly reset)	__% (e.g. 5%) improvement between each annual mean rate	Operator voluntary hazard report rate (e.g. per 1 000 FH)	Consideration	Consideration
			Operator DGR incident report rate (e.g. per 1 000 FH)	Average + 1/2/3 SD (annual or 2 yearly reset)	__% (e.g. 5%) improvement between each annual mean rate

FIGURE 3.17 Air operators' SMS safety performance indicators suggested by the ICAO.

Source: ICAO, 2013b, p. 5 APP 6–4.[51]

Most of these indicators focus on operational events to check whether the safety risk controls in place are effective, including the mitigations put in place that were supposed to reduce unacceptable risks. A few others refer to other aspects like the rate of audit findings or the number of voluntary hazard report rates.

Besides operations, other aspects are monitored by SPIs. The UK CAA, for example, provides an indicative list of safety indicators (developed in 2013) in a "safety performance indicator template" that includes performance indicators such as not only the number of major incidents but also the number of internal audits, the number of safety newsletters issued or the number of safety surveys.[52]

The UK CAA makes it explicit in the template document that "the suggested objectives are an example only. Organisations should set objectives that are relevant to their particular type of operation."

In practice, such quantitative indicators on operations, reporting, or the SMS process, such as the number of safety committee meetings or audits, are commonly used by aviation organizations.

Beyond the definition of SPIs, the monitoring of safety performance relies on the collection and analysis of data. A key aspect is the existence and actual implementation of a reporting system for hazards and events that guarantees reporter confidentiality. The other sources of data are diverse. They may include safety studies on dedicated issues, safety data analysis based on report events or systematic data monitoring, safety surveys, safety audits, findings, and recommendations from safety investigations.

However, the range of aspects covered and the way they are addressed in these various data sources varies dramatically from one service provider to another. For example, the way safety events are analyzed varies in practice from a quick analysis limited to immediate causes such as errors and non-compliance, whereas other organizations will look for root causes and distal factors that may have contributed to the event.

This diversity of safety model and safety culture among aviation organizations is also reflected in the chosen SPIs. From purely quantitative indicators on operational events and safety meetings to more qualitative aspects that reach beyond operations and the formal SMS process, there is a wide range of approaches to the safety performance monitoring element and, more generally, to the safety assurance pillar. For example, the Xyz Air example adopts a simplistic view of safety culture by reducing it down to four SPIs in, namely, the number of safety meetings of the safety board, number of safety drills, number of operative personnel trained, and number of technical personnel trained. CASA provides in its booklet an approach to determine a safety culture index based on safety culture surveys addressing a much more complete range of aspects related to the five key ingredients of safety culture according to Reason (1997), CASA (2014c, pp. 40–41).

Likewise, safety dashboards used by some organizations to report the safety performance of the organization and its evolution to the accountable executive (and possibly other senior managers [e.g., the SRB]) vary from one organization to another (for those that are using a dashboard).

In practice, beyond the SPIs and their assessment that can be compiled by the SMS manager, the monitoring of safety performance translates into the various safety committees (at least the SRB and SAGs for large organizations), meetings where the various indicators are discussed, and corrective actions decided and assigned. How the structures (e.g., SRB, SAGs, safety management group, and safety management group or personnel) work varies from one organization to another. As an illustration, for a major airline with a fleet of more than a hundred aircraft, the SRB may meet twice a year and the SAGs (usually one per type of activity) four times a year, whereas other safety meetings, organized and chaired by the safety manager—for example, on the analysis of reported events (and more generally, feedback from experience)—may meet weekly. However, beyond the written traces and the organizational structure (defined as part of the safety organization), a major part of the safety assurance pillar, but also the less easy to access, is the content of the discussions and decisions taking place during the safety committee meetings.

The Management of Change

The supporting material on the management of change largely remains at the level of principles as to what are the changes to be considered. The practical support, when developed in supporting resources, focuses on how to address a change.

Indeed, the UK CAA states in its guidance document, "Organisations should define the types of changes that would require a formal management of change process" (CAA, 2014, CAP 795, p. 20) and provide a "management of change template" with a set of questions to characterize the change (e.g., "Define the major components or activities of the change," "Who does the change affect?" or "What is the impact of the change?") and to assess the associated risk (e.g., "What is the issue?" "What could happen as a result?").[53]

Whether in the guidance document for air carriers or for airports (including in the sample SMS manual provided in the Appendix), the US FAA does not refer explicitly to change management as part of safety assurance as a dedicated area.

The fictitious case of Xyz Air developed by EASA focuses its monitoring on

> its operations and the environment to assure that it recognizes changes in the operational environment that could signal the emergence of new and unmitigated hazards, and for degradation in operational processes, facilities, equipment conditions, or human performance that could reduce the effectiveness of existing safety risk controls.
>
> **(EASA, Xyz Air SMS Manual, p. 42)**

The self-assessment checklist provided by the Australian CASA in the booklet on safety assurance illustrates a broader understanding of all the aspects that could play a role on safety and thus ultimately affect safety performance (Figure 3.18). It embraces a number of managerial and organizational aspects:

- Management structure
- Management and corporate stability
- Financial stability of the organization
- Management selection and training
- Workforce
- Relationship with the regulatory authority

Self-assessment checklist

You can use the following self-assessment checklist to identify administrative, operational and other processes, and training requirements, that might indicate safety hazards. You can then focus attention on those issues posing a possible safety risk.

Management and organisation

Management structure

1. Does the organisation have a formal safety policy and written safety objectives?
2. Are the corporate safety policies and objectives adequately disseminated throughout the organisation? Is there visible senior management support for these safety policies?
3. Does the organisation have a safety department or a designated safety manager (SM)?
4. Is this department or SM effective?
5. Does the SM report directly to the accountable manager?
6. Does the organisation support the periodic publication of a safety report or newsletter?
7. Does the organisation distribute safety reports or newsletters from other sources?
8. Is there a formal system for regular communication of safety information between management and employees?
9. Are there periodic safety meetings?
10. Does the organisation participate in industry safety activities and initiatives?
11. Does the organisation formally investigate incidents and accidents? Are the results of these investigations disseminated to managers and operational personnel?
12. Does the organisation have a confidential, non-punitive, hazard and incident reporting program?
13. Does the organisation maintain an incident database?
14. Is the incident database routinely analysed to determine trends?
15. Does the organisation operate a flight data analysis (FDA) program?
16. Does the organisation operate a line operations safety audit (LOSA) program?
17. Does the organisation do safety studies?
18. Does the organisation use outside sources to do safety reviews or audits?
19. Does the organisation seek input from aircraft manufacturers' product support groups?

Management and corporate stability

1. Have there been significant or frequent changes in ownership or senior management within the past three years?
2. Have there been significant or frequent changes in the leadership of operational divisions within the past three years?
3. Have any managers of operational divisions resigned because of disputes about safety matters, operating procedures or practices?
4. Are safety-related technological advances implemented before they are directed by regulatory requirement, i.e. is the organisation proactive in using technology to meet safety objectives?

Financial stability of the organisation

1. Has the organisation recently experienced financial instability, a merger, an acquisition or other major reorganisation?
2. Was consideration given to safety matters during and following the period of instability, merger, acquisition or reorganisation?

FIGURE 3.18 Extract from the CASA booklet on safety assurance illustrating the self-assessment checklist.

Source: CASA, 2014d, p. 18.

In practice, the range of changes considered as possibly affecting safety is not similar across aviation organizations. They are indicative of the organization's safety model or understanding of what might contribute (positively or negatively) to safety and safety culture.

Continuous Improvement of the SMS

Although this element is called "continuous improvement of the SMS" in ICAO Annex 19, thus in all CAAs legislations and guidance documents, it is sometimes interpreted as the continuous improvement of safety performance.

For example, the US FAA guidance document for air carriers states, "The final step within SA is continuous improvement. This process is designed to ensure that you are correcting substandard safety performance identified during the safety performance assessment in order to continuously improve safety performance" (FAA, 2015, AC 120-92B, p. 45). In this document, continuous improvement is mainly about making sure, through the safety assurance activities described previously, that risk controls are actually put in place wherever they were meant to be.

The UK CAA similarly states, "The organisation should continually seek to improve their safety performance" (CAA, 2014, CAP 795, p. 20). As for the Australian CASA, it refers instead to the continuous improvement of safety management.

Overall, the guidance provided suggests that the continuous improvement is ensured through safety committee reviews, safety surveys, safety audits, and the like.

In practice, many SMS manuals don't include a specific section regarding the continuous improvement of SMS, nor are there specific activities conducted in the framework of this element.

Pillar 4: Safety Promotion

Per ICAO Annex 19, this pillar includes the two following elements:

4.1 Training and education
4.2 Safety communication

The fourth and last pillar of SMS requirements in aviation explicitly connects the SMS with safety culture in the ICAO SMM:

> Safety promotion encourages a positive safety culture and helps achieve the service provider's safety objectives through the combination of technical competence that is continually enhanced through training and education, effective communication, and information-sharing. Senior management provides the leadership to promote the safety culture throughout an organization.
>
> **(ICAO, 2018b, pp. 9–25)**

The ICAO SMM dedicates a whole chapter to safety culture, defined as "an expression of how safety is perceived, valued and prioritized by management and employees in an organization" (ICAO, 2018b, p. 3–1). Considerations on safety culture, way beyond training to safety or safety communication, are developed:

> Staff concerns about workload, job security and access to training are associated with significant change in organizations and can have a negative impact on safety culture. The degree to which staff feel involved in the development of change and understand their role in the process will also influence the safety culture.
>
> **(ICAO, 2018b, pp. 3–3)**

The existence of diverse "safety cultures" even within the same organization is discussed, whether national, professional, ethnical, or other types of culture, as well as the challenge for managers "to promote a common understanding of safety and each individual's role in its effectiveness" (ibid.).

The central role of safety culture is not only acknowledged but also emphasized: "Safety culture is arguably the single most important influence on the management of safety" (ICAO, 2018b, p. 3–1).

However, beyond the general discourse on safety culture, the translation of the safety promotion pillar (supposed to foster a positive safety culture) into supporting material and practical guidance reduces the reach of safety culture to safety training and communication.

Training and Education

Supporting material on training and education is relatively limited and leaves significant leeway to organizations as to how to implement it.

The section on training and education developed in UK CAA CAP 795 is limited to general objectives customized to the role of the staff. For example, "operational staff should have an understanding of the organisation's safety policy and the principles and processes of the organisation's SMS"; additionally, "managers and supervisors should understand the safety process, hazard identification, risk management and the management of change" and even "organisational safety standards, safety assurance and the regulatory requirements for their organization" for senior managers (CAA, 2014, CAP 795, p. 21). As for the accountable manager, she/he "should have an awareness of SMS roles and responsibilities, safety policy, safety culture, SMS standards and safety assurance" (CAA, 2014, CAP 795, p. 21).

The US FAA guidance material for airports distinguishes the training to be delivered between individuals and managers with operational roles under SMS who should receive a specialized SMS training and all other individuals who should receive safety awareness/orientation (possibly through written communication such as brochures only) (FAA, 2016, DRAFT AC 150/5200-37A).

The most detailed guidance is that provided in the Australian CASA booklet of safety promotion (CASA, 2014e). It includes an example of content for the safety training of various personnel, as illustrated in Figure 3.19.

In practice, the training delivered in relation to the SMS varies dramatically from one organization to another and within the same organization, from one type of personnel to another. For the personnel without a specific role related to the implementation of the SMS, it may range from written information posted somewhere on the information system to awareness sessions, including discussions and debates around safety.

For managers, it may also range from written information posted somewhere on the information system to facilitated sessions on the role of management in safety culture, including practical exercises, just culture including recognizing positive behaviors and so on, beyond a mere explanation of the mechanics of the SMS.

For safety managers, the same diversity exists between a low-cost training limited to brief e-learnings on the SMS pillars, risk analysis, and the other aspects of the mechanics of SMS, to extensive trainings, such as advanced masters.

TYPE OF TRAINING	SAMPLE CONTENT
Initial employee induction Initial SMS implementation	» Safety philosophy, safety policies and safety standards including: - approach to 'safety culture' - not apportioning blame - difference between acceptable and unacceptable behaviour - internal safety investigation policy and procedures » The content of the SMS and rationale for it » Importance of complying with the safety policy and with the standard operating procedures that form part of the SMS » Organisational roles and responsibilities of staff in relation to safety » Organisational safety record, including areas of systemic weakness » Procedures for hazard reporting » Organisational safety management programs (e.g. reporting system, internal audit program, line operations safety audit [LOSA] etc.) » Requirement for ongoing internal assessment of organisational safety performance (e.g. employee surveys, safety audits and assessments) » Reporting ATSB and WH&S reportable matters, hazardous events and potential hazards » Lines of communication for safety matters » Feedback and communication methods for disseminating safety information » Safety awards programs (if applicable) » Safety promotion and information dissemination » Emergency response.
Management awareness	» The manager's role in shaping safety/reporting culture, including a 'just culture' » The safety risk management process » Managers' responsibilities and accountabilities for safety » Managers' legal liabilities under CASA legislation » Safety committee's risk assessment/root cause analysis » Safety promotion/communication and information dissemination.
Safety-critical personnel	» Procedures for reportable matters » Specific safety initiatives, such as: threat and error management (TEM), crew resource management (CRM), approach and landing accident reduction (ALAR), maintenance error decision aid (MEDA), and line operations safety audit (LOSA) » Seasonal safety hazards and procedures (weather-related operations etc.) » Emergency procedures/response » Current/recent safety situations » Safety promotion/communication and information dissemination.
Safety specialist	» Monitoring safety performance » Conducting risk assessments » Managing the safety information system (database) » Performing safety audits » Understanding the role of human performance in accident causation and prevention » Operation of the SMS » Investigation of reportable matters and hazardous events » Crisis management and emergency response planning » Safety promotion/communication methods » Communication skills » Computer skills such as word-processing, use of spreadsheets and database management.

FIGURE 3.19 Sample content for safety training provided by CASA for each personnel profile.

Source: CASA, 2014e, p. 3.

Maintaining records of the trainings delivered within an organization is part of the SMS manager's responsibilities. In practice, the number of trainings delivered is also commonly used as an SPI, without any consideration of the content of the training. Therefore, it is not uncommon to observe organizations paying more attention to the number of trainings than to the relevance of the content.

Safety Communication

Safety communication is commonly understood as the main "vector" of safety culture. Just as for safety training, the guidance material is rather general and leaves significant leeway to organizations to implement safety communication requirements.

For example, the UK CAA guidance material on safety communication includes a list of types of communication as well as a list of its objectives (see CAA, 2014, CAP 795, pp. 21–22). The type of communication mentioned is safety policies and procedures, newsletters, safety bulletins and notices, presentations, websites and emails, and informal workplace meetings between staff and the accountable manager or senior managers. As for the objectives of safety communication, they combine very generic and wide ones such as "ensure that all staff are fully aware of the SMS and the organisation's safety culture," and more specific ones such as "explain why safety procedures are introduced or changed" (CAA, 2014, CAP 795, pp. 21–22).

Likewise, the US FAA guidance material for airports is rather light on safety communication. Although the main message is "Communicating safety information is essential to promoting a safety culture," the documentation only mentions four items to consider "when deciding what information to communicate," respectively related to the awareness of SMS and its initiatives, the communication of safety-critical information, the explanations for new or revised safety procedures, and the communications records to be maintained (see FAA, 2016, DRAFT AC 150/5200-37A, p. 47).

The Australian CASA booklet on safety promotion goes further into detail as to what safety communication is about and how it translates (see Figure 3.20). For example, the safety communication that happens during shift handovers (and the conditions for it to happen, especially sufficient time) is explicitly considered, although it is not necessarily part of what is commonly understood by safety professionals as "safety-critical information." Likewise, the fact that communication should be two-way is emphasized, as well as the importance of discussing events with the workforce, rather than just disseminating information about events, to make organizational learning possible.

The differences in the mindset of the supporting material illustrate the diversity of practical implementations of the safety communication element across aviation organizations. In practice, safety communication ranges from a minimum communication of the mandatory aspects with record-keeping as a key concern to a broader approach encompassing more dimensions of safety culture.

3.3 CONCLUSION: FROM RELIANCE ON SMS STANDARDS TO ORGANIZATIONAL REFLEXIVITY—A MATTER OF PHILOSOPHY

Although the SMS pillars and constitutive elements are common to all aviation organizations, there are notable variations in the approach to supporting aviation

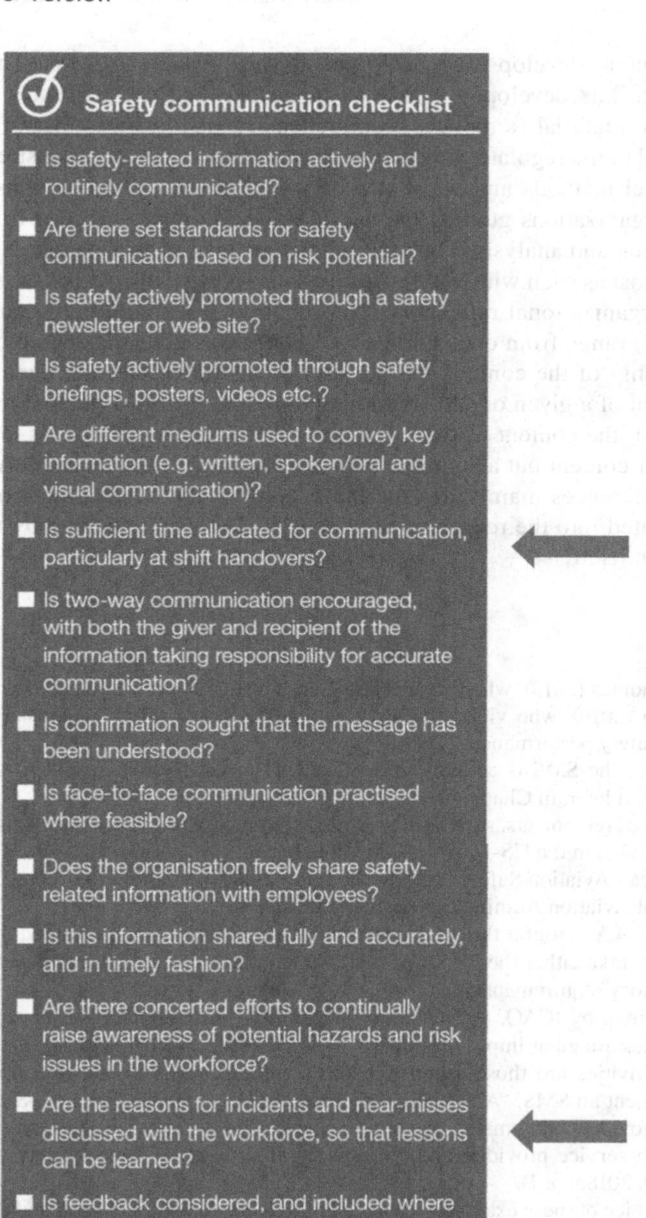

FIGURE 3.20 Safety communication checklist provided by CASA.

Source: CASA, 2014e, p. 8.

organizations to develop their SMS and, in turn, in actual SMSs. The guidance and resource kits developed by CAAs range between two extremes: (1) almost ready-to-use material (templates, comprehensive risk analysis, list of indicators) that respond to the regulatory requirement of having an SMS; (2) upstream support on high-level methods and ways of working with some singular illustrations on fictitious organizations guiding the organization in how to proceed to develop its own reflection and analysis. The first option can be used as a standard to be implemented almost as such with very little customization, if any. As for the second one, it invites organizational reflexivity. The required resources (both qualitative and quantitative) range from one extreme to the other, as does the degree of relevance and ownership of the content and mindset of the SMS. Overall, even though the SMS manual of a given organization may look like the SMS manual from another organization, the content of the safety policy, risk analysis, or SPIs may vary not only in their content but also in the way in which they were enacted. At the origin of these differences mainly lies the philosophy of the development of the SMS, both translated into the resources provided by CAAs and adopted by the aviation organization itself.

NOTES

1. See Thomas (2012), who considers SMS as a delegation of responsibility, and Rae and Provan (2019), who views the SMS as administrative safety somewhat decorrelated from safety performance.
2. Whether the SMS is actually up to these promises in practice will be analyzed and discussed later, in Chapter 4.
3. Recent developments, such as the major evolution of ATM—namely, SESAR in Europe or NextGen in the US—are not addressed here.
4. European Aviation Safety Agency, the European CAA.
5. Federal Aviation Administration, the US CAA.
6. Many CAAs around the world, especially in countries with a limited aviation background, take either the EASA or the FAA regulation as a source to develop their own regulatory requirements.
7. As defined by ICAO, a state safety program is "an integrated set of regulations and activities aimed at improving safety" (ICAO, 2018b, p. viii). Among these regulations and activities are those requiring each service provider or aviation organization to implement an SMS. "A State's safety programme, combined with the SMSs of its service providers, systematically addresses safety risks, improves the safety performance of each service provider, and collectively, improves the State's safety performance" (ICAO, 2018b, p. 1).
8. The choice of these examples, especially the UK CAA, the US FAA and the Australian CASA was made because the guidance material and resources they are providing to support service developers in the development of their own SMS are very diverse and illustrate different philosophies and methodologies to implement an SMS.
9. Before the creation of the Annex 19 in 2013, fully dedicated to safety management, safety related requirements were disseminated in existing annexes. Safety management program recommendations were introduced in 2001 for Air Traffic Services and Certified Aerodromes in Annexes 11 and 14, respectively. SMS recommendations were introduced in 2006 in the same annexes for these two same service providers. They were then introduced in 2007 for air operators and approved maintenance organizations

(applicability in 2009) in Annex 6, for approved training organizations in 2009 (applicability in 2010) in Annex 1, and finally for aircraft and manufacturing organizations in 2009 (applicability in 2013) in Annex 8.

10. ICAO is a UN organization. As such, it cannot issue regulations or rules, but the SARPs presented in their annexes are the basis on which states or regions develop their regulations.
11. For EU countries, SMS requirements are defined in EU regulations, more specifically in the European Aviation Safety Programme. However, national authorities have their own state safety program. They also develop their own guidance material and other resources to support their service providers in the implementation of SMS.
12. Annex to the report to the European Parliament and the Council—The European Aviation Safety Programme COM (2015) 599 final (https://eur-lex.europa.eu/resource.html?uri=cellar:f0a0e4cd-9ce8-11e5-8781-01aa75ed71a1.0020.02/DOC_2&format=PDF).
13. As stated by the UK CAA on the webpage dedicated to bow-ties, bow-ties can be used to fulfil certain SMS goals in an appropriate, efficient, and consistent way (link provided in endnote 23).
14. See www.caa.co.uk/Safety-Initiatives-and-Resources/Working-with-industry/Bowtie.
15. See www.govinfo.gov/content/pkg/FR-2015-01-08/pdf/2015-00143.pdf.
16. FAA Order 8000.369 (A for the first edition, B for the second).
17. Safety regulations documentation in Australia is structured as follows: the Civil Aviation Safety Regulations 1998 (CASR) and Manuals of Standards (MOSs) are legislations. Advisory Circulars (AC) and Acceptable Means of Compliance and Guidance Material (AMC/GM) provide guidance associated with the legislations.
18. As of March 2020, the SMS legislation on SMS varies from one type of organization to another. A project is ongoing to move towards harmonized legislation and guidance material in the form of a new Part 5 of Civil Aviation Safety Regulations, a new Manual of Standards and supporting material.
19. Considering the current dissemination of SMS requirements in the Civil Aviation Safety Regulations and Manual of Standards, they are not all mentioned here. However, they can be found on the CASA website using this link: www.casa.gov.au/safety-management/safety-management-systems/legislation-and-guidance.
20. Statements such as "It describes some options for complying with the Civil Aviation Regulations 1988 (CAR 1988)" or "This publication is only advisory but it gives a CASA preferred method for complying with the Civil Aviation Regulations 1988 (CAR 1988)" can be found at the beginning of CAAPs to remind their status.
21. Additional practical content to support service providers in the development of their SMS.
22. For the sake of readability, we will not repeat guidance material and supporting resources to refer to all the material with no legislation status developed by authorities to support their service providers in developing and maintaining their SMS, but this is the reference material used as a basis for the whole section.
23. Examples of this ready-to-use material will be provided in the following sections dedicated to each of the four SMS pillars.
24. Respectively, bush aviation and training, a fictitious air transport operator, a small family-owned charter business operating from a regional airport, Outback Maintenance Services, and a small maintenance organization based at a regional airport.
25. Some variations may be noted in the structure of the documentation (e.g., whereas the UK CAA and Australian CASA also develop the SMS documentation aspect as part of the first pillar, the FAA dedicated a whole part to safety documentation). However, the point of this analysis is not to focus on the structure itself but rather on the content and its underlying characteristics.

26. As mentioned earlier, this fourth element will not be addressed since it refers to the post-accident phase and thus belongs to the crisis management area which is not within the scope addressed in this document. Besides, it is not applicable to all aviation organizations (e.g., aircraft manufacturers).
27. See www.caa.co.uk/Safety-initiatives-and-resources/Working-with-industry/Safety-management-systems/Safety-management-systems.
28. The case studies taken as fictitious examples in the CASA booklets are small organizations.
29. Large organizations may have a dedicated safety department led by the head of safety management.
30. That is including, for example, investigating incidents and accidents, providing safety advice to management and staff, or reporting to the executive accountable the SMS performance and any need for improvement.
31. In large organizations, the safety manager is supported by other safety management personnel.
32. See www.caa.co.uk/Safety-initiatives-and-resources/Working-with-industry/Safety-management-systems/Safety-management-systems.
33. See www.ukfsc.co.uk/files/SMS%20Material/EHEST%20SMS%20Toolkit/EHSIT%20SMS%20Manual%20-%20Final%20Draft%20Dec%202011.pdf.
34. Reproduced with the permission of ICAO.
35. Reproduced with the permission of ICAO.
36. For example, data from accidents, incidents, flight data monitoring, voluntary and confidential reporting systems, open hazard reporting systems, LOSA (Line Operations Safety Audit) style normal operation assessments, safety surveys, change management processes, workshops with subject matter experts, and safety committee meetings.
37. The UK CAA bow-tie templates will be extensively described and discussed in the following subsection.
38. "Organisations may use barrier models such as bow-tie for their risk management process" (CAA, CAP 795, p. 13).
39. See www.caa.co.uk/Safety-initiatives-and-resources/Working-with-industry/Bowtie/Bowtie-templates/How-were-the-bowtie-templates-created-/.
40. See www.caa.co.uk/Safety-initiatives-and-resources/Working-with-industry/Bowtie/Bowtie-templates.
41. The top event is the moment when the system is destabilized.
42. See www.caa.co.uk/Safety-initiatives-and-resources/Working-with-industry/Bowtie/Bowtie-templates/Bowtie-document-library.
43. UK CAA website: www.caa.co.uk/Safety-initiatives-and-resources/Working-with-industry/Bowtie/Bowtie-elements/Escalation-factors.
44. See www.caa.co.uk/Safety-initiatives-and-resources/Working-with-industry/Bowtie/Bowtie-templates/Bowtie-document-library.
45. See www.caa.co.uk/Safety-initiatives-and-resources/Working-with-industry/Bowtie/Bowtie-templates/Bowtie-document-library.
46. Whether the fact that complete bow-tie templates are easily available played a role in this choice or not is worthy of investigation.
47. The supporting material developed by CAAs is quite open and diverse.
48. See www.ukfsc.co.uk/files/SMS%20Material/EHEST%20SMS%20Toolkit/EHSIT%20SMS%20Manual%20-%20Final%20Draft%20Dec%202011.pdf.
49. This manual served as guidance material for CAAs and was the reference ICAO SMM until the issuance of the fourth version in 2018. Interestingly, there are no longer any detailed examples of SPIs provided in the fourth version issued in 2018. However, they constitute ready-to-use material that some aviation organizations commonly rely on.

50. For example, the engine in-flight shutdown incident rate is obviously specific to air operators, whereas the quarterly runway foreign object/debris hazard report rate is specific to aerodrome operators.
51. See www.caa.co.uk/Safety-initiatives-and-resources/Working-with-industry/Safety-management-systems/Safety-management-systems.
52. Reproduced with the permission of ICAO.
53. See www.caa.co.uk/Safety-initiatives-and-resources/Working-with-industry/Safety-management-systems/Safety-management-systems.

REFERENCES

Alizadeh, S. S., & Moshashaei, P. (2015). The Bowtie method in safety management system: A literature review. *Scientific Journal of Review*, *4*(9), 133–138.

Bird, F. E. J., & Loftus, R. G. (1976). *Loss control management*. Institute Press (A Division of International Loss Control Institute).

Bouwmans, I., Weijnen, M. P., & Gheorghe, A. (2006). Infrastructures at risk. In: *Critical infrastructures at risk* (pp. 19–36). Springer.

CAA. (2014). *Safety Management Systems (SMS) guidance for organisations*. Document CAP 795. CAA. https://publicapps.caa.co.uk/modalapplication.aspx?appid=11&mode=detail&id=6616

Cacciabue, P. C., Cassani, M., Licata, V., Oddone, I., & Ottomaniello, A. (2015). A practical approach to assess risk in aviation domains for safety management systems. *Cognition, Technology & Work*, *17*(2), 249–267.

Civil Aviation Safety Authority (CASA). (2014a). SMS for aviation: A practical guide. In: *Safety management systems basics* (2nd ed.). CASA. www.casa.gov.au/files/2014-sms-book1-safety-management-system-basicspdf

Civil Aviation Safety Authority (CASA). (2014b). SMS for aviation: A practical guide. In: *Safety policy and objectives* (2nd ed.). CASA. www.casa.gov.au/files/2014-sms-book2-safety-policy-objectivespdf

Civil Aviation Safety Authority (CASA). (2014c). SMS for aviation: A practical guide. In: *Safety risk management* (2nd ed.). CASA. www.casa.gov.au/files/2014-sms-book3-safety-risk-managementpdf

Civil Aviation Safety Authority (CASA). (2014d). SMS for aviation: A practical guide. In: *Safety assurance* (2nd ed.). CASA. www.casa.gov.au/files/2014-sms-book4-safety-assurancepdf

Civil Aviation Safety Authority (CASA). (2014e). SMS for aviation: A practical guide. In: *Safety promotion* (2nd ed.). CASA. www.casa.gov.au/files/2014-sms-book5-safety-promotionpdf

Dekker, S. (2019). *Foundations of safety science: A century of understanding accidents and disasters*. Routledge.

Dien, Y., & Dechy, N. (2013), Les risques organisationnels des "organisations fragmentées." *Les entretiens du risques*.

Federal Aviation Authority (FAA). (2015). *FAA AC_120-92B, safety management systems for aviation service providers*. www.faa.gov/documentLibrary/media/Advisory_Circular/AC_120-92B.pdf

Federal Aviation Authority (FAA). (2016). *DRAFT AC 150/5200-37A, safety management systems for airports*. www.faa.gov/documentlibrary/media/advisory_circular/draft-150-5200-37a.pdf

Gheorghe, A. V., Masera, M., Weijnen, M., & De Vries, L. (2006). *Critical infrastructures at Risk: Securing the European electric power system*. Springer.

Grote, G. (2012). Safety management in different high-risk domains: All the same? *Safety Science*, *50*(10), 1983–1992.
Hale, A. R. (2005). *Safety management, what do we know, what do we believe we know, and what do we overlook*. Safety Science Group Delft University of Technology.
Hale, A. R., Heming, B. H. J., Carthey, J., & Kirwan, B. (1997). Modelling of safety management systems. *Safety Science*, *26*(1–2), 121–140.
Hale, A. R., & Hovden, J. (1998). Management and culture: The third age of safety: A review of approaches to organizational aspects of safety, health and environment. *Occupational Injury: Risk, Prevention and Intervention*, 129–165.
Harris, D., & Stanton, N. A. (2010). Aviation as a system of systems. *Ergonomics*, *53*(2), 145–148.
Heinrich, H. W. (1931). *Industrial accident prevention: A scientific approach*. McGraw-Hill.
Hovden, J. (1998). The ambiguity of contents and results in the Norwegian internal control of safety, health and environment reform. *Reliability Engineering & System Safety*, *60*(2), 133–141.
Hovden, J., & Tinmannsvik, R. K. (1990). Internal control: A strategy for occupational safety and health: Experiences from Norway. *Journal of Occupational Accidents*, *12*(1–3), 21–30.
ICAO. (2013a). *Annex 19: Safety management* (1st ed.). ICAO.
ICAO. (2013b). *Safety management manual (SMM)* (3rd ed.). Doc 9859. ICAO.
ICAO. (2016). *Annex 19: Safety management* (2nd ed.). ICAO.
ICAO. (2018a). *Annex 19: Safety management* (3rd ed.). ICAO.
ICAO. (2018b). *Safety management manual (SMM)* (4th ed.). ICAO.
Itabu Issa Sadiki, M. (2017). La sécurité aérienne en Afrique: la communication autistique au sein du collectif sécuritaire de l'aéronautique civile congolaise. Sociologie. Université Panthéon-Sorbonne-Paris I.
Kontogiannis, T., Leva, M. C., & Balfe, N. (2017). Total safety management: Principles, processes and methods. *Safety Science*, *100*, 128–142.
Kysor, H. D. (1973). Safety management system: Part I: he design of a system. *National Safety News*, *108*, 98–102.
Li, Y., & Guldenmund, F. W. (2018). Safety management systems: A broad overview of the literature. *Safety Science*, *103*(2018), 94–123.
Maier, M. W. (1998). Architecting principles for system of systems. *Systems Engineering*, *1*, 267–284.
Maurino, D. (2017). *Why SMS: An introduction and overview of safety management systems*. OECD. www.itf-oecd.org/sites/default/files/why-sms.pdf
Merritt, A., & Maurino, D. (2004). Cross-cultural factors in aviation safety. In: *Cultural ergonomics* (pp. 147–181). Emerald Group Publishing Limited.
Rae, A., & Provan, D. (2019). Safety work versus the safety of work. *Safety Science*, *111*, 119–127.
Reason, J. (1997). *Managing the risks of organizational accidents*. Ashgate.
Swuste, P., Groeneweg, J., Van Gulijk, C., Zwaard, W., & Lemkowitz, S. (2018). Safety management systems from Three Mile Island to Piper Alpha, a review in English and Dutch literature for the period 1979 to 1988. *Safety Science*, *107*, 224–244.
Thomas, M. J. (2012). *A systematic review of the effectiveness of safety management systems* (No. AR-2011-148). Australian Transport Safety Bureau.
Vaughan, D. (1996). *The challenger launch decision: Risky technology, culture, and deviance at NASA*. University of Chicago Press.
Voss, B. (2012, May). SMS reconsidered. *Aerosafety World*, p. 1.

4 What Does the SMS Actually Do, and Is It Up to Its Safety Promises?

This chapter proposes an alternative version of the SMS. Beyond the declared promises of the SMS, the SMS users' perceptions of the approach are presented as a starting point for a critical analysis of what the SMS actually can and cannot do from a safety perspective. On the basis of the detailed description of what it consists of provided in the previous chapter, we explore the limitations of the SMS. The analysis highlights a number of shortcomings of the SMS as a safety dispositive. In particular, we point out the conceptual, methodological, and practical limitations. Conceptually, the SMS may allow for identifying and managing operational risks, especially those generic enough to be modeled. However, it turns out to leave apart a number of other risks, especially those potentially induced by the organization itself, and to mainly focus on real-time risks. Besides, SMS, safety, and risk are confused to a certain extent. Safety is considered an isolated stake that can be managed independently from the other organizational issues. As for the most used methodology supporting the risk analysis, it does not reflect the risk itself but a proxy of it that can be misleading. The limitations are even greater when the SMS implementation itself is minimal. The SMS as practiced turns out to be less promising than announced as a safety approach.

4.1 HOW THE SMS ACTUAL USERS PERCEIVE IT

In the second edition of their reference book on SMSs in aviation published in 2015, Stolzer and Goglia write,

> Regulatory authorities, safety experts, and industry leaders still believe that SMS represents the future of safety management in the aviation industry. SMSs provide organizations with a powerful framework of safety philosophy, tools, and methodologies that improve their ability to understand, construct, and manage proactive safety systems.
>
> **(Stolzer & Goglia, 2015, p. 15)**

However, as Hovden identified for internal control in its time, stakeholders' perceptions about SMS are diverse. Beyond the enthusiasm for the SMS reported by the authors, some practitioners express more skeptical views.

At a high level, SMS is sometimes perceived by practitioners as the ultimate proceduralization of all safety aspects in aviation rather than a real shift from a compliance-based to a performance-based approach to safety (Pelegrin, 2013). As stated by an SMS manager on one of his tasks associated with the implementation

DOI: 10.1201/9781003307167-4

of the SMS, "Completing an SMS maturity survey takes days. I don't want to see another questionnaire. It's a nightmare. That's the practical perspective. The SMS questionnaire is an administrative overhead" (safety manager from a European aviation organization, June 20, 2017).[1]

As for the promise to put safety on top managers' agendas, it seems that the expectation is not always fulfilled in the way in which it was imagined. "The topic is on their agenda, but the concern is not always. There is a 30 min safety discussion at the very beginning of the board meetings, but after that, they start talking real business" (mixed experience 2, diverse industries, January 22, 2019).

Even in mature organizations where safety culture is rather well developed, what SMS can deliver is not always perceived as a direct result of safety performance as such.

> Performance is not the outcome of SMS. Performance is the outcome of a lot of things an organization is doing. Reality is a set of competing pressures. We need to manage the risks the best we can. We're here to put ourselves (safety people) out of business. As a safety manager, you are the conscience of the organization.
>
> **(Safety manager from a European aviation organization, June 20, 2017)**[2]

Or as stated by one of the interviewees, "The SMS doesn't replace anything like an accident prevention program, safety culture. . . . It gives a way to prioritize/allocate resources. The safety programs will do this by combining everything" (regulator 2, aviation, September 25, 2018). Or said differently, "We haven't waited for the SMS to manage safety" (safety manager of a European aviation organization, May 5, 2015).

Similar divergences in perceptions were observed earlier regarding the implementation of internal control. Based on an extensive analysis of actors' perceptions, Hovden concluded in 1990 that the way internal control was perceived by industrials and authorities ranged from a bureaucratic approach imposed by authorities to a breakthrough in improving safety (Hovden, 1990).

With the benefit of hindsight, it seems that something went wrong between the intention of the SMS, its safety promises, and its actual implementation. The report is bitter: "For the past 20–30 years, the main focus has been on the packaging of safety management. The basic content hasn't changed" (mixed experience 4, oil and gas, September 17, 2020). "This SMS wave led to a lost decade of reflection on what makes safety. It reinforced the religion of compliance, and its deviation sins, indicators" (mixed experience 2, diverse industries, January 22, 2019).

Why is the SMS apparently not fully up to its safety promises? The following sections provide insights into this question by developing conceptual, methodological, and practical limitations of the SMS.

4.2 WHAT THE SMS ACTUALLY DOES: AN INSIDERS' PERSPECTIVE

Answering the question of what the SMS actually does is not simple and straightforward. Indeed, even within the same industry, the diversity of implementation across companies is huge (Li & Guldenmund, 2018; Kaspers et al., 2017). The guidance material provided by CAAs (see Chapter 3) already provides a flavor of the range of interpretations and translations of the four SMS pillars as illustrated earlier by different organizations. Nevertheless, this section will start with a review and discussion of major generic assumptions underlying the SMS, whatever form it takes in practice. Some methodological choices regarding risk analysis considered as the key step change proposed by the SMS will also be discussed. Finally, some practical aspects involved in the SMS will also be analyzed and discussed with respect to their implications.

4.2.1 Conceptual Shortcuts and Assumptions

SMS and Safety Performance

The former ambition of the SMS is to enhance the safety performance, or at the international level, as stated by ICAO, to "assist States in managing aviation safety risks, with the objective of continuously reducing the number of aviation accidents and incidents" (ICAO, 2013, p. A7). Such an ambition involves an implicit assumption whereby having a good SMS leads to having a good safety performance. The UK CAA also puts forward the relationship between SMS and safety performance by making the following link: "Accidents occur as a result of safety management systems failing."[3] Yet the relationship between the two seems far from being obvious, according to the few authors who have analyzed it. Thomas, in his systematic review of the effectiveness of the SMS, underlines "a lack of consistency . . . with respect to relationships between SMS and safety performance" (Thomas, 2012, p. 24). Following their study in aviation, Kaspers et al. conclude that "there is limited empirical evidence about the relationship between Safety Management System (SMS) processes and safety outcomes" (Kaspers et al., 2017, p. 9) and that "there is a limited value of linear thinking followed by the industry—that is, "the more you do with an SMS the higher the safety performance" (Kaspers et al., 2017, p. 9). Finally, considering the SMS as a proxy for safety performance turns out to be an empirical shortcut, partly due to conceptual shortcuts.

Defining Safety Performance Indicators

Monitoring safety performance, part of the third pillar of SMS (safety assurance), involves, as suggested by the existing guidance material illustrated in the previous section, the definition of safety performance indicators (SPIs). Defining safety indicators or SPIs has been the topic of passionate debate (e.g., Reiman & Pietikäinen, 2012; Kjellén, 2009; Wreathall, 2009). Among them is the distinction between lagging and leading indicators—that is, between indicators that change values respectively after and before the level of safety of the organization has changed (Kjellén, 2009, p. 486). Whereas lagging indicators focus on the negative and the outcome, thus being reactive, leading indicators focus on the positive and the processes in place to control the risk, thus being proactive (Reiman & Pietikäinen, 2012). Classic examples of lagging indicators are the number of fatal accidents or the number of significant incidents. Conversely, leading indicators can translate into the number of safety reviews performed per year or safety culture surveys. Reiman and Pietikäinen (2012) go further by distinguishing between drive indicators (indicating the potential of the organization to achieve safety), monitor indicators (indicating the culture of the organization), and outcome indicators (measuring the outcome of the sociotechnical system) (Reiman & Pietikäinen, 2012, p. 1996).

In an ethnographic study of flight safety investigators, Macrae (2009) highlights that flight safety investigators themselves, when trying to identify new risks based on near-misses analysis, use ignorance (that is, doubts and suspicions) as a proxy indicator of risk. They decide to further investigate events where they find a gap or uncertainty in their own knowledge in relation to an organizational event. Macrae concludes that there cannot be any a priori threshold to decide whether an event is critical or not and requires further investigation since it depends on the organization's investigators' current state of knowledge, which evolves with time (p. 287).

Eventually, defining indicators that actually give indications on the level of safety is far from obvious. Thus, how do we define the goals of the SMS? Yet this question is essential. Indeed, the goals

> drive the performance through identification of discrepancies between actual performance and goals through feedback loops. If the goal is defined badly, if it does not measure what it is supposed to measure, then the system can't possibly produce a desirable result. Systems tend to produce exactly and only what you ask them to produce.
>
> **(Meadows, 2008, p. 138)**

As a consequence, a number of organizations tend to manage indicators rather than actual performance or phenomena underlying the measure of indicators. This tendency can be supported by the behavioral theory of the firm in which organizational decision-makers act as satisficers rather than maximizers (Simon, 1957). Therefore, setting goals and indicator thresholds (i.e., organizational aspirations) allows them to know when organizational performance is satisfactory (March & Simon, 1958). "Goals transform continuous measures of performance into dichotomous variables. Performance falls either above the decision maker's aspiration (and is considered a success) or below the aspiration level (and is considered a failure)" (Madsen, 2013, p. 767). Yet a number of cautions and warnings have been expressed by research on indicators and their use. As summarized by Reiman and Pietikäinen (2012), "Safety management should be about managing the sociotechnical system, not about managing and optimizing certain indicators" (Reiman & Pietikäinen, 2012, p. 1999). Interestingly, ICAO puts forward such a warning with respect to the assessment of safety culture, that conversely does not appear as a monitored regulatory requirement as such: "Also, scoring safety culture maturity can have unintended consequences by inadvertently encouraging the organization to strive to achieve the 'right' score, rather than working together to understand and improve the safety culture" (ICAO, 2018, p. 3–8). The face value of indicators is challenged at a philosophical level by Hacking (2001), who questions the objectivity and truth of scientific objects. Hacking considers that a type of reasoning is more than a set of techniques aimed at highlighting new types of facts. It also creates its own truth criteria. In other words, it is self-justifying. Such criticisms are also shared at a practical level, especially by SMS managers. As stated by one of them, "These indicators are necessary evils for they need to be reported to the regulator but don't tell anything about the safety of performance" (safety manager of an aviation organization, June 20, 2017).[4] Safety indicators provide a simplistic view of what actually contributes to safe operations, as metaphorically illustrated in Figure 4.1.

FIGURE 4.1 Illustration of the difference between "reality" embraced in its complexity and through indicators.

Source: Inspired by Ursus Wehrli's *The Art of Clean Up*.

Safety and Risk Management

From a conceptual point of view, the SMS relies on the following definition of safety: "The state in which risks associated with aviation activities, related to, or in direct support of the operation of aircraft, are reduced and controlled to an acceptable level" (ICAO, 2016, p. 1–2). Yet the definition of safety is not unique. Indeed, safety is sometimes seen as the absence of negative events such as accidents like it used to be historically or is still in the literature on the profit-safety link (Madsen, 2013).

With such a definition, nothing is explicitly said as to what safety relies on, but compliance with regulatory requirements is understood as the means to achieve it. Social sciences understand safety rather as an emergent property of the functioning of the entire system that is neither a state nor the result of a process but the result of dynamic ways of working and adapting to a changing environment and context, allowing the management of the unexpected beyond what has been anticipated, pre-organized, and proceduralized (Bieder & Bourrier, 2013; Weick & Sutcliffe, 2015). The definition of safety underlying the SMS refers instead to risks. In aviation, safety is defined in the SMS documentation as the state in which risks are reduced and controlled to an acceptable level. Such a definition involves a number of implicit assumptions and limitations.

One of these implicit assumptions is related to the scope of risk management. Indeed, although rarely made explicit, the scope of risk management is limited to the domains of the known and knowable (Desroches et al., 2003). Therefore, risk management does not address the infinite domain of the unknown-unknows, or the aspect of uncertainty in a broad sense that involves the events or phenomena no one has ever anticipated or imagined (Bieder, 2017), or the black swans as defined by Taleb (2007), although they may lead to catastrophic consequences. As recalled by Dupuy and Grinbaum (2005), uncertainty is not limited to epistemic uncertainty but also encompasses objective uncertainty in complex systems.

Furthermore, by relying on risk management, the SMS puts the focus on formal knowledge and generic models reflecting the system as a whole and makes any deviation from safety management "by design" or operations "as imagined" a source of unsafety, despite the acknowledgment that departing from rules and procedures is sometimes what is needed to operate safely, especially when uncertainty is high (Bourrier, 2017; Grote, 2012). It also thereby overlooks the tacit experiential knowledge, the necessary real-time adaptations that allow for managing operations in their diversity and complexity (Schulman et al., 2004; Bieder & Bourrier, 2013; Hollnagel, 2014). Indeed, the SMS is limited to a managerial level perspective relying on models of the system and its functioning (i.e., a simplification of reality that overlooks the specificities of singular situations). As noted by Schulman et al. (2004), although reliability professionals play a key role in navigating the middle ground and making interconnected network operations reliable, thus safe in the HRO perspective, they are absent from formal risk analysis.

Likewise, all the organizational considerations acknowledged, especially in the 1990s, as playing a key role in safety (Bourrier, 2017) are oversimplified. In the framework of risk analysis, all this body of knowledge is somehow ignored and comes down to considering as hazards "organizational factors, such as those related to the recruitment, training and retention of personnel, the compatibility of production and safety goals, the allocation of resources, operating pressures and the corporate safety culture" (ICAO, 2009, pp. 5–17).

In addition, by limiting the scope of the risk analysis to safety events, thereby disregarding the other performance objectives and stakes of the organization, some risk reduction barriers may undermine the achievement of these other objectives (e.g., productivity). In such cases, they are more likely to be bypassed by operators to overcome this induced reduction in efficiency in other areas than safety.

Risk Management

Although the 1950s and 1960s were decades of great progress in terms of the development of methods to manage the risk of new technologies, some theoretical and methodological challenges remain unaddressed regarding risk management, including in the framework of the SMS.

Risks from What End: Confusion between Causes and Consequences

The first challenge is scoping out the risks to be considered. According to the definition of risk proposed in the ISO 31000 norm on risk management, risk is defined as the "effect of uncertainty on objectives" where "an effect is a deviation from the expected (positive and/or negative)"; "objectives can have different aspects (such as financial, health and safety, and environmental goals) and can apply at different levels (such as strategic, organization-wide, project, product and process)" (ISO, 2009, p. 1, 2018). Therefore, risk can encompass a variety of aspects, such as financial, safety, security, social, political, and environmental. Yet when one refers to environmental risks, confusion remains as to whether "environmental" refers to the nature of the hazards or that of the consequences.[5,6] Likewise, when addressing safety risks, a company can refer to safety either as a hazard likely to lead to consequences, especially on the company itself, like bankruptcy or loss of reputation (e.g., the fatal TWA 800 or Malaysia MH370 accidents had a negative impact on the reputation of the airlines, at least for some time), or as consequences (i.e., potential harm to human beings, the environment or property/equipment). In the first case, the catastrophic consequences are the enterprise's bankruptcy or the end of its activity, and safety is one hazard among others. Indeed, there are many more risks beyond an accident that can "kill" a company, like having dissatisfied customers, not finding markets anymore, or not being able to fund necessary investments. In this landscape, the occurrence of an accident with critical consequences is just one misfortune among others. Thus, safety can be understood as an art of trade-off (Amalberti, 2017). In the second case, there is a range of aspects that can affect the operations of the company (whether directly or not) and lead to an accident with safety consequences.

What Focus for the Risk Analysis?

Most risk analysis methods and practices in high-risk industries still focus on operational processes and aspects despite more and more evidence of the interrelation between all the dimensions an organization has to handle (e.g., financial, safety, environment, quality) and the associated risks. As an illustration, accident investigation reports often mention, beyond operational elements (e.g., technical failures, human errors), more indirect contributing factors at an organizational level. For example, in the case of the Colgan Air accident, the pilots' wage policy was identified by the NTSB as a factor contributing to pilots' fatigue. Indeed, considering the high cost of living in the Newark area and the low wages crew members received from the airline, a number of pilots were unable to afford living in this area and had to do several-hour commutes to take on their duties (NTSB, 2009). Nevertheless, up to now, very few methods propose a broad range of generic hazards to be considered in the a priori analysis of risks with potential safety consequences (e.g., the preliminary risk analysis, recently extended to the global risk analysis [Desroches et al., 2009, 2016]), and their use remains an exception.[7] Reason himself, although pushing the theory of the organizational accident in the late 1980s to the early 1990s, challenged "too wide a net" in the search for contributing factors, claiming that moving away from local contributing factors had a limited return on risk management (Reason, 1999). Most risk analyses performed in the framework of the SMS focus on operational risks (Dekker, 2019; CAA, 2014)—that is, activities performed at the sharp end. The UK CAA even explicitly recommends disregarding transverse organizational factors, such as fatigue, as escalation factors in bow-ties and developing a specific bow-tie dedicated to fatigue. By doing so, organizational aspects that could appear as common modes of operations are disconnected from operations. More generally, integrating organizations into risk analysis presents a number of challenges (Pettersen Gould, 2020; Le Coze, 2005) that are not yet overcome, even more so by techniques such as bow-ties. As stated by Petersen Gould,

> If risk analysis is to broaden its scope and hazardous organizations are to have risk policies to minimize risks while at the same time being able to deal with the unexpected and rare crisis, one must start by acknowledging that there are no simple solutions for paralleling these.
>
> **(Pettersen Gould, 2020, p. 7)**

Acceptability

Another lasting challenge in risk management is that of the definition of "acceptable." The whole concept of risk management in high-risk industries revolves around the definition of a threshold between acceptable and unacceptable risks, defined based on the combination of the severity of consequences and probability of occurrence. Although most risk management approaches call for the definition of severity and probability scales, there is very little theoretical work supporting the definition of these scales. To account for uncertainty, Aven (2014) suggests an adjustment procedure for determining risk acceptance, taking into account the strength of knowledge used as a basis for the assessment of the risk probability but does not examine how to define the various severity levels. Desroches et al. (2003, 2009) proposes a characterization of the five levels of the severity scale, where *catastrophic* corresponds to "safety failed or loss of the system" (i.e., irreversible consequences on safety or on the system's integrity), *hazardous* to "safety or system's integrity degraded," *major* to "mission failed" with no safety impact, *minor* to "mission degraded" with no safety impact, and *insignificant* to "no significant consequences" (Desroches, 2009, p. 60). By defining major and minor consequences with respect to the system's mission, Desroches illustrates that safety and the main purposes/objectives of the system are interrelated. Indeed, if, in order to reduce a safety risk, the mission of the system is jeopardized with a certain likelihood, the risk will remain unacceptable.

Determining the "acceptability threshold" also raises some questions and conveys some implicit assumptions. As raised by Heimann (1997, 2005) and reinforced by Bourrier (2005), "Which risks has the organization chosen to be 'acceptable'? (type I, loss of installation and life; type II, waste of resources). Does everybody agree with this choice? Are there any dissenting voices?" (Bourrier, 2005, p. 103). More generally, is it even possible to talk about the risks that an organization chooses to be acceptable when a significant variety of risks are considered by the various actors of an organization and these risks may vary depending on the role these actors play in a given point in time? At similar levels within a high-risk organization, a safety officer/manager may have views that differ from that of the finance officer/manager on the "system" and the risks considered. The gap in scopes may be even bigger at different organizational levels or between operational and support functions where the focus is different (Bieder & Callari, 2020).

Risk to Whom/What System Is Considered?

Modeling a system involved in risk management with multiple actors assumes that risk management is understood in a similar way by all of them. Beyond the challenges addressed before, the various actors may consider different scales/systems, thus different safety risks, and consider different levels of acceptability. Indeed, risks only make sense and can be defined with respect to a system. As stated in the FAA mission cited earlier, the system they consider is the aerospace system, whereas an airline, for example, would rather consider its own operations, which is not at all the same system. For a given organization (e.g., an airline), ending its activity is a catastrophic consequence on its own scale. Yet at a wider scale, that of the air transport system, as long as the function/mission of the given organization is sustained (i.e., aircraft are still operated even though by different operators), there is no catastrophic consequence. In a similar way, the risks managed by a pilot are focused on the flights she/he is performing, whereas the airline as an organization manages the risks of its overall fleet operations (and possibly of its interactions with other partners of its alliance, if any). Not being able to perform her/his flight may be considered by a pilot as a mission failed, whereas "sacrificing" this specific flight to the benefit of the management of the overall airline network may be considered far less severe by the airline as an organization. The "no rule, no use" phenomenon highlighted by Fucks & Dien (2013) illustrates that, in some organizations, some employees may even have a tendency to ask for rules and comply with them—that is, to manage the risk of being accountable, thus blamed for taking an initiative, rather than managing the risks of the operational situation itself. In other words, they balance an immediate risk to the "company system," that of the operational situation, and a longer-term risk to "themselves as a system," that of losing their job. What are the risks that are managed by the various actors is not an easy question with an easy answer.

What Time Horizon(s) to Consider?

Accident investigations in aviation but also in other hazardous activities often highlight the contribution of decisions and/or actions that were taken way before the event. One of the most commonly acknowledged examples in aviation is the erosion of pilots' manual flying skills with cockpit automation (Casner et al., 2014). However, one can doubt that this hazard and associated risks were considered, thus anticipated decades back when modern cockpits were developed in the 1980s.

Aviation combines safety risks that can emerge at a wide range of time horizons: from very short-term to very long-term. Real-time operational conditions like a bird strike call for an immediate response. At the other end of the spectrum, an aircraft design choice may induce a safety hazard decades later, as was observed with cockpit automation choices made in the 1980s and progressively leading to the decrease of pilots' manual skills, still required today in some situations. Macrae (2019) introduces the notion of time in his proposed characterization of three different scales of organizational activities by distinguishing between situated, structural, and systemic. Situated refers to the operational situation itself, structural emerges in the close monitoring of operations, whereas systemic emerges in the oversight of system structure and interaction. Schulman (2020) proposes to enlarge timeframes to encompass slow-motion failures and intergenerational risk and achieve a higher-resolution additive understanding of organizational and managerial factors in safety and reliability. So far, though, time had been a remarkably absent dimension in risk analysis and safety management methods and approaches. The time horizon (or horizons) considered remains implicit and thus may vary from one analyst to another. Possible long-term consequences are commonly disregarded without this limitation being made explicit.

Until now, considerations of time in safety/risk management have rather consisted in looking back in time to identify contributing factors such as organizational or regulatory ones in retrospective analysis (Reason, 1997; Rasmussen, 1997) or by considering "changes" over the past years and analyzing their potential impact on safety as in the safety evaluation model proposed by Le Coze (2011, p. 135). Potential consequences of high-risk activities and associated risk management strategies and actions are rarely anticipated at different time horizons in the future. Intergenerational risks (for example, nuclear waste disposal) and necessary long-lasting actions to manage such risks over decades or possibly centuries are typically never addressed in a priori or proactive risk analysis (La Porte, 2020). Yet "the world peeps, squawks, bangs, and thunders at many frequencies all at once" (Meadows, 2008, p. 104). There is no one-time horizon or delay that is more important than another in the absolute. Important delays depend on the purpose of the discussion (just as appropriate boundaries around one's picture of a system). If one is worrying about the oscillations that take weeks, one probably does not have to think about delays that take minutes or years. What is a significant delay usually depends on which set of frequencies one is trying to understand (ibid.). The difficulty is that each actor has his/her purpose(s), thus, his/her own important delay(s), whereas this diversity is often overlooked and ignored.

Can Managing Safety Be Achieved by a Dedicated System?

The SMS relies on a socio-technical model of control where the management controls that operations at the sharp end are actually happening the way they were imagined to. It conveys the underlying postulate that there can be a system in charge of ensuring risk or safety management. However, can a socio-technical system be engineered to manage safety as a main purpose? If we consider the various aspects considered in these models of control within companies, they involve resources, work instructions, company policies, and work actions. All these elements are part of a broader context of the company and also contribute to other stakes and objectives of the company. If we consider, for example, the standard operating procedures used as a basis for the risk analysis for an airline, they are developed based on a combination of especially operational and safety aspects. Ultimately, "the very essence of a company . . . is first and foremost to produce a product or a service to exist, survive, develop itself and blossom" (Amalberti, 2017, p. 25). Safety management is one of the sub-purposes of a company, but all the actors of a company contributing to safety also contribute to the overall company objectives. Likewise, at the level of governments or regulators, managing safety cannot be considered the sole objective.

Consequently, the regulations and standards produced by regulators already embody some trade-offs allowing the agencies to manage their coexisting objectives. For example, the safety-related regulatory requirements are such that they allow aviation to further develop rather than putting an exclusive focus on safety, possibly to the detriment of economic feasibility or sustainability. The multiple purposes are also part of the interactions/negotiations between regulators and regulatees (Baram & Lindoe, 2018; Jasanoff, 2003). At a different level, the type of regulation philosophy and the authorities' internal resources, both qualitatively and quantitatively, are adjusted to one another. The same applies to the number and depth of inspections and the resources available (Dupré et al., 2014).

Most of the organizations and elements of the so-called socio-technical system involved in safety management also form part of other systems having other purposes. Whatever the system, defined by Meadows (2008) as "an interconnected set of elements that is coherently organized in a way that achieves something" (Meadows, 2008, p. 11), its purpose is often the most crucial determinant of its behavior and "a change in purpose changes the system profoundly, even if every elements and interconnections remain the same" (p. 17). Such changes were introduced when airlines added to their former objective of ensuring safe transport of passengers and/or cargo, other purposes such as making money (especially for former state-owned airlines being privatized), extending their market, and providing low-cost operations. Thus, how do we describe a socio-technical system of control, a system involved in risk or safety management, when its elements contribute in parallel to a variety of purposes evolving with time?

In practice, most large companies have developed an increasing number of management systems: quality management system "owned" by the quality division, SMS "owned" by the safety division, or security management system "owned" by the security division or an integrated management system, all of which apply to all employees throughout the organization. The objective of these management systems is to have a footprint on the organization itself and the way it operates. As illustrated by Cochoy and De Terssac (2009), "quality approaches support the rationalization of productive activities by extending the action of organizing to all the actors and to all the stages of the development and production of a service or a product" (p. 5). This rationalization, although not focused on safety, interferes with the way that safety can be managed, all the more so when procedures are then used as an instrument of control of the activity, thus as the reference to a normative approach to safety. More generally, within an organization, the same actors involved in the socio-technical system of control of safety are also involved in other systems of management of other stakes, such as quality or security, thus pursuing different purposes.

Finally, what matters to make a system successful is to keep "sub-purposes and overall system purposes in harmony" (Meadows, 2008, p. 16). Whether one can then legitimately draw the boundaries of the systems contributing to the sub-purposes and define them is debatable. "There are no separate systems. The world is a continuum. Where to draw a boundary around a system depends on the purpose of the discussion—the question we want to ask" (Meadows, 2008, p. 97). However, problems arise when we forget that we have artificially created them (ibid.). The introduction of the notion of the SMS somehow artificially creates a boundary around the "system" supposedly in charge of managing safety for the purpose of studying/enhancing safety. But the boundaries are artificial as well as isolating safety from the rest.

As a summary, the very notion of the SMS raises a number of questions:

- How do we account for the fragmentation of organizations and thus define the boundaries of organizations?
- How do we account for the multiple organizations involved in aviation operations?
- Does it make sense to consider a system involved in safety management when its elements also contribute to other purposes and change behaviors when these purposes evolve?
- Does it make sense to isolate safety from other stakes when they are so intricate, including in the interactions between the various actors?

A Limited View of Management

The driver to move away from past approaches focusing on technical aspects and human factors (limited to individuals) is a commonly shared acknowledgment synthesized by Turner et al. (1989): "The human factors that undermine safe operations are of a more subtle kind. Typically they are complex, multifaceted and rooted in the social, managerial and organizational properties of the particular socio-technical system concerned" (Turner et al., 1989, p. 3). Whereas the SMS focuses on the allocation of resources to the most critical safety issues, a sort of risk-informed resource allocation approach, the social, managerial, and organizational aspects encompass much more aspects. Leadership or relationship with employees or deference to expertise, for example, are part of the factors identified as contributing to safety. As stated by Andéol-Aussage et al., "le domaine du management ne se limite pas à celui de la gestion" (Andéol-Aussage et al., 2018, p. 24).[8] In other words, there is more to management than modeled and quantified aspects. As illustrated by one of the interviewees,

> There is an area here, human relationships, psychological relationship between the manager and employees who are supposed to be there to provide the information needed for improving things: how can you possibly ever regulate that? A manager who fails with that will not learn much organizationally. A manager who refuses to acknowledge his own mistakes to workers and that sort of things. There is a whole bunch of things related to organizational learning. How do you get into this level of details? You cannot do it through prescriptive standards. You cannot prescribe that management shall be curious of his employees and offer him a cup of coffee. You can't do it through prescriptive, you can't do it through performance-based either.
>
> **(Academic 4, diverse industries, June 29, 2020)**

By aiming at being a standardized approach, the SMS relies on a simplified and even simplistic definition of management, overlooking all the subtleties that were pointed out as contributing to safety. In the end, there is a shift between the initial idea of the role played by organization and management in safety and the management of safety as understood in the SMS that mainly comes down to a safety administration (or even sometimes accountancy) approach. Or maybe should we say a drift from "management *and* safety," management here being understood in a broad sense that includes human and qualitative context-specific aspects, to the "management *of* safety," management being understood in an administrative control way relying on generic models and quantification.

4.2.2 Methodological Limitations

As mentioned earlier, the main novelty introduced by the SMS is the use of a risk management approach to anticipate and proactively manage safety risks. Therefore, this section will focus on the methodological cornerstone of the SMS, namely, the risk analysis. Although there is no single risk analysis method imposed by SMS regulatory frameworks, bow-ties are among the most commonly used technique in the SMS (Alizadeh & Moshashaei, 2015; Cacciabue et al., 2015). Beyond the general conceptual limitations of risk analysis mentioned earlier, bow-ties convey specific methodological choices and limitations that are worth exploring to understand what the SMS actually does.

As stated by the UK CAA on their website, "Bowtie is one of many barrier risk models available to assist the identification and management of risk and it is this particular model we have found (and are still finding) useful."[9] However, not all risk analysis methods are similar in terms of their conceptual foundations, their scope of validity, the resources (both qualitative and quantitative) required to implement them, or the relevance of their outcomes. Without getting back to the origins of the bow-tie technique, it combines fault-trees and event-trees techniques and comes down to a cause-consequence diagram. It starts with the identification of a top event, usually based on feedback from experience (thus rather reactive), and goes backward to the potential causes and forward to the potential consequences.

Graphically, it proposes a simple representation of so-called safety barriers that prevent either a hazard from developing into a top event (prevention part) or a top event from developing into safety-related consequences (recovery part), as illustrated in Figure 4.2.

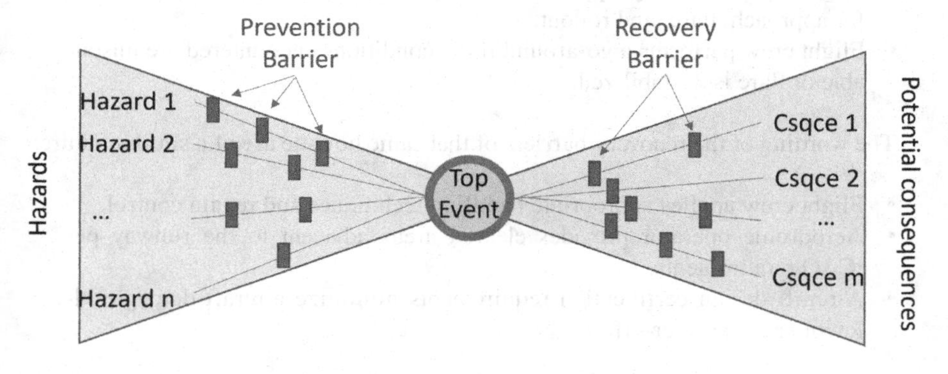

FIGURE 4.2 Generic format of a bow-tie diagram.

However, what does it actually tell us about risks (even restricted to operational risks)? ICAO defines a safety risk as "the predicted probability and severity of the consequences or outcomes of a hazard" in its Annex 19 on safety management (ICAO, 2013, 2016).

As such, a risk relates a hazard to certain consequences having a certain severity. A closer look at the bow-tie templates provided by the UK CAA shows that the so-called consequences are worded in such an ambiguous way that the severity cannot always be determined. For example, the consequence of the bow-tie on runway excursion (2.2), illustrated in Figure 3.13 and Figure 3.14, is worded so: "Aircraft veers off side of the runway colliding with structures, obstacles or terrain resulting in injuries/fatalities." First of all, fatalities and injuries are not usually categorized with the same severity. Besides, the probability of an aircraft veering off the side of the runway and actually colliding with structures, obstacles, or terrain varies from one airport and even runway to another, just as the probability of this collision leading to injuries or to fatalities. To summarize, the consequences, as they are worded, already include a certain scenario following the runway excursion itself.

Additionally, bow-ties focus on safety by design or safety engineering (Acfield & Weaver, 2012) or safety as imagined to build on the distinction proposed by Hollnagel (2014) between "work as imagined" and "work as done." The safety barriers modeled in bow-ties usually reflect the compliance with design specifications for technical components and with operational procedures by individuals at the sharp end. As an illustration, the wording of the prevention barriers that can be found in the first scenario of the "runway excursion" bow-tie illustrated earlier, taken from the UK CAA website, reads as follows:[10]

- Aerodrome operator maintains runway to ICAO friction standards.
- Aerodrome operator adheres to runway clearance procedure (e.g., contamination removal).
- Aerodrome operator communicates "slippery when wet" conditions to flight crew.
- ATCO provides accurate meteorological and runway state information to flight crew.
- Flight crew correctly applies crosswind-handling techniques/procedures for approach, flare, and rollout.
- Flight crew performs a go-around if the conditions encountered are unsuitable or flare is destabilized.

The wording of the recovery barriers of that same bow-tie are of a similar nature:

- Flight crew applies appropriate handling techniques and regain control.
- Aerodrome operator provides cleared areas adjacent to the runway per ICAO requirements.
- Aircraft design certification requirements minimize aircraft damage following runway veer-offs.

- Flight crew and cabin crew perform aircraft evacuation.
- Aerodrome emergency response plan implementation.

Although the SMS pretends to reach beyond a mere compliance-based approach to safety, when relying on bow-tie as a technique to model and assess risks, it somehow transfers the compliance base from regulatory requirements to aviation organizations' rules or operational procedures or ICAO requirements. In that respect, it may not be an adequate way of "managing the unexpected," although its existence is recognized (Maurino, 2017).

Furthermore, even great defenders of bow-tie diagrams recognize, "It is not possible to make bow-tie diagrams for all existing hazards. There are simply too many hazards, and it will take too much time to make as many bow-tie diagrams" (Sankar & Siddiqui, 2016, p. 213). In addition, bow-ties tend to oversimplify complex situations (Popov et al., 2016). Indeed, the modeling proposed by the bow-tie technique is linear and disregards interactions between the so-called safety barriers. As mentioned earlier, organizational aspects are reduced to escalating factors when modeled. Last, bow-tie diagrams are a static representation. All these characteristics somehow contradict the initial reason why the SMS was introduced in air transport, namely, the acknowledgment of the increasing complexity of aviation.

Besides, observed risk management real practices in organizations using bow-tie diagrams challenge even further the use of the bow-tie technique as a proxy of risk analysis. Indeed, in some organizations, the author observed the number of times a top event was tagged as "occurred" (through the analysis of reported events) being used as a risk indicator. Yet the choice of the top event is subjective.[11] In practice, top events are of very different intrinsic criticalities in the sense that the robustness of the recovery varies dramatically from one event to another. For example, let us consider two bow-ties (extracted from the UK CAA list of bow-tie templates) (see Figure 4.3).

Without going into technical and operational details, if we only consider the recovery maneuvers required from the flight crew to recover from the top event, "applying appropriate handling technique to regain control" (as stated in the first bow-tie) or "performing upset recovery procedure" (as stated in the second one) are not equally robust. The second procedure is an emergency one to get out of an emergency situation such as a stall. In other words, an occurrence of the first top event is "further away" from the aircraft veering off the side of the runway than the second one from an unrecovered loss of control in-flight.

Additionally, although the consequences of both top events can be similar, namely, fatalities, the number of runway excursion accidents leading to fatalities does not compare with the number of loss-of-control in-flight accidents leading to fatalities. While LOC-I (loss of control in-flight) accidents account for 33% of the fatal accidents and 12% of the hull losses, RE (runway excursion) accidents account for 16% of the fatal accidents and 36% of the hull losses (Airbus, 2020). To put it bluntly, runway excursions "kill" more rarely than the loss of control in-flight. A risk being defined with respect to the probability of a hazard leading to

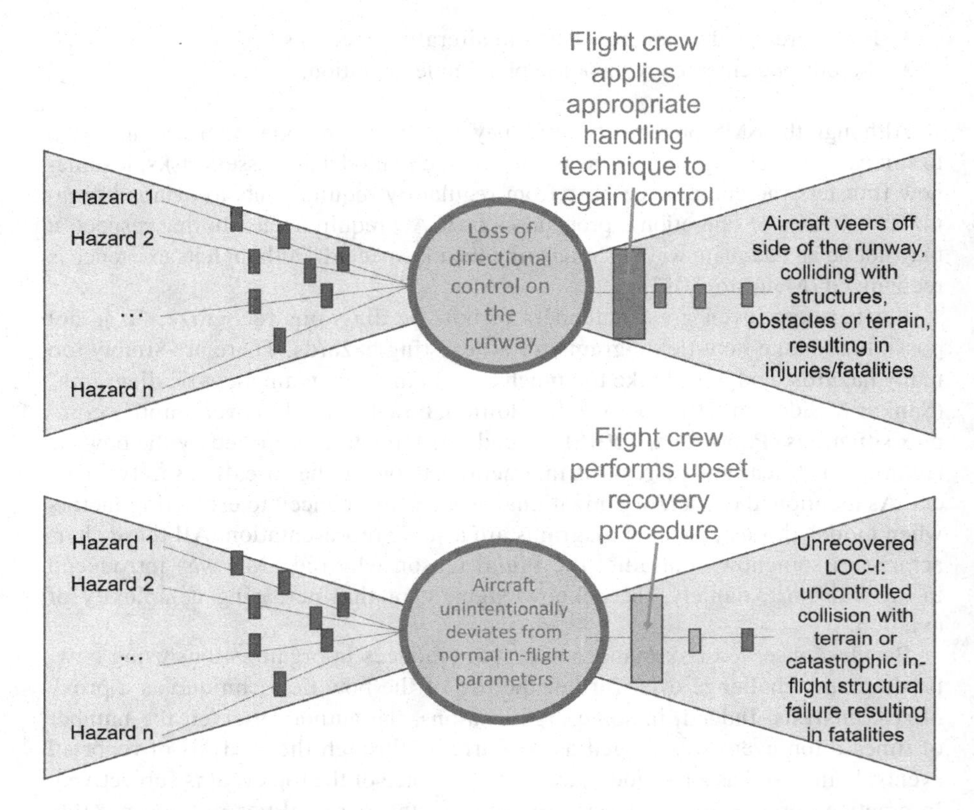

FIGURE 4.3 Simplified bow-ties from the respective "runway excursion" and "loss of control in-flight" templates.

Source: Adapted from the UK CAA website as of March 2020.

consequences of a certain severity, the occurrence of the top event does not tell us much about the risk since the top event is arbitrarily positioned along several accident scenarios that would need to be considered in their entirety, from a hazard to the consequences, to say something about the risk. The list of bow-ties cannot be prioritized since each bow-tie mixes several scenarios, making their use extremely limited when it comes to allocating resources to the reinforcement of the most critical safety barriers.

Although bow-tie diagrams are a powerful graphical tool to communicate, especially to illustrate that an accident can have different causes and consequences, they are not really a risk analysis tool as such.

4.2.3 Practical Limitations

Qualitative and Quantitative Resources

Developing and maintaining an SMS requires a certain amount of resources, both qualitative and quantitative, that are not always available within aviation organizations. A common example is the knowledge required to perform a risk analysis, whatever the method. Therefore, aviation organizations that cannot easily afford to properly develop their own SMS use alternative strategies (the same applies in other high-risk industries as well). Some of them rely on consultants to take care of their SMS (Almklov et al., 2014; Almklov, 2018; Voss, 2012); others replicate the guidance material put together by regulators, adding their name and logo without any permeation, the result sometimes being a combination of both. As William R. Voss put it when he was president and CEO of the Flight Safety Foundation,

> Back when the international standards for SMS were signed out at ICAO, we all knew we were going to launch a new industry full of consultants. We also knew that all these consultants couldn't possibly know much about the subject and would be forced to regurgitate the ICAO guidance material that was being put out . . . All of those predictions have come true.
>
> **(Voss, 2012, p. 1)**

Consultants, for the sake of their budget, but also for their lack of familiarity with the organization's specificities as outsiders, tend to "sell" as generic a material as possible (Almklov, 2018). As for the guidance material made available by regulatory authorities, it is also generic enough to be "applicable" to any organization. In both cases, the SMS as implemented by these organizations is missing the point of customizing the approach and the safety enhancement measures, one of the promises of the SMS.

Even so-called mature organizations developing their own SMS, actually putting resources to customize it to their own reality, complain about the number of resources needed to manage the bureaucratic part of the SMS. Among the top risks within his organization, a safety manager mentions, "Safety staff is overworked to manage the SMS" (safety manager from a European aviation organization, November 23, 2017).

Managing Indicators More Than Safety, Complying with SMS Requirements More Than Managing Safety

Besides missing the point of customizing solutions, relying on the existing guidance material or consultant ready-to-use approaches based on acceptable means of compliance comes down to complying with the SMS regulatory requirements. Thereby, SMS managers manage SPIs without making sense of them, considering them as a proxy of safety performance.

In the end, the SMS efficiency safety-wise turns out to be limited in practice. The experience of the introduction of internal control in Norway in the 1980s seems to recur: "A paradox for the IC [internal control] reform so far is that it has contributed more to making enterprises with good SHE [safety, health, and environment] records better, than in improving the SHE conditions in ignorant enterprises" (Hovden, 1998).

4.3 WHAT MAKES SAFETY SO SPECIFIC?

Considering the significant potential for harm to people in high-risk industries, safety can be seen as the absolute frontier defended by a moral sense. If "safety first" is a frequent motto in many high-risk organizations or safety regulatory agencies, one can wonder how it translates into practice on a daily basis. Is an ethical dimension guiding decisions and actions when safety is involved, or is safety more a constraint than anything else, or maybe one organizational stake among many others, with no specificity?

The "moral sense of safety" historically leading people to do what is right (in a moral sense) for safety is observed by Fucks and Dien (2013) to be declining due to organizational trends contributing to developing the belief that actions taken beyond the borders of prescriptions are too risky. This organizational evolution, meant to reinforce a sense of responsibility at work, has the opposite effect and leads, in fact, to a decline of responsibility and vigilance, thus to "a decline in the value of safety—the well-known 'safety first'" (p. 34). Wachter (2011), coming from a different angle from the proceduralization of safety, shares the conclusion that apart from HSE (health, safety, and environment) managers who want to do the right thing (in a moral sense), the organization instead wants to make a profit and adopts a cost-benefit approach to safety rather than an ethical one. He even dismissed the possibility of envisaging an ethical approach to safety as too foreign to current collective consciousness and experience and suffering from irresolvable tensions between ethics and capitalism. Indeed, conversely to a regulatory approach achieved through enforceable rules, an ethical approach deals with values and is self-regulated.

If safety is no longer considered the absolute frontier, it is for most systems perceived, like environment, as a constraint on how the mission can be achieved (Leveson, 2009).

> There are always multiple goals and constraints for any system—the challenge in engineering and management decision making is to make tradeoffs among multiple requirements and constraints when the designs and operational procedures for best achieving the requirements conflict with the constraints.
>
> **(Ibid., p. 236)**

However, by considering safety as a constraint, its share of attention and share of voice is limited by design. With a broader perspective of the purpose of an organization at a societal level (i.e., beyond shareholders and even employees) and with a broader range of time horizons, safety could be seen as belonging to the list of the main purposes of an organization or system.

Critically, other research has shown that regulation was not the only driver to safety; neither is a quantitative cost-benefit approach. Indeed, safety can also be seen as part of the social license that pushes companies to reach beyond compliance with regulatory requirements and maintain a reputation capital. Even if the benefits of doing so are difficult to assess, Gunningham et al. (2004) identify this as a condition to stay competitive, especially since it reduces uncertainty by managing the risk of adverse social campaigns and lubricates the regulatory process by providing access to political actors and with regulatory flexibility. The concepts of "social license" or "social contract" introduce another key stakeholder in the governance and management of safety—the civil society. Whereas the civil society is absent as such from the models of risk control inspired by systems engineering, the public nevertheless turns out to be an actor as such in at least two respects. Due to the increasing defiance toward scientists, experts, and institutions, they represent a protest power that can damage the reputation of a company/organization/institution (Millstone & Van Zwanenberg, 2000). On the other hand, they hold significant knowledge and information; thus, involving them is acknowledged as beneficial to safety and its management (Jasanoff, 2003; Gunningham et al., 2004; Bastide, 2018). One of the reasons is the high degree of uncertainty attached to safety. If uncertainty is not exclusively due to the lack of knowledge, traditional risk governance based on risk assessment relies on a limited and simplified knowledge: limited because of relying on a limited number of scientific experts but also simplified because of relying on generic models (Jasanoff, 2003).

4.4 CONCLUSION: BEYOND THE FACE VALUE OF THE SMS

As an approach to enhance safety, the SMS turns out to have a number of limitations. Although it may help in identifying major and relatively stable issues, all the dynamic aspects and contingencies go through the SMS net without being spotted or addressed. They are even ignored in a way since, as a system pretending to be inclusive, they are left apart, unaddressed. Depending on the mindset in which the SMS is implemented, from a real opportunity to think about actual safety concerns within the organization and collectively find ways to make progress to a mere cut-and-paste exercise of existing generic material, the outcome might even be somehow detrimental to safety by giving the illusion that safety is managed since there is an SMS. Eventually, with this feedback after years of implementation, one could wonder why such an approach is still considered the cornerstone of aviation safety regulation. Criticisms were and still are formulated by practitioners and academics, especially on the bureaucratic character of the SMS (Hudson, 2001; Dekker, 2014). The SMS is considered as safety work and opposed to the safety of work—that is, operational safety (Dekker, 2019). Yet the SMS is still standing, and there might be some good reasons for it. Even though its main "official" purpose is not completely fulfilled, why is the SMS still widely used as a safety management approach?

Rae and Provan (2019) recently proposed reaching beyond the opposition between the safety of work (or operational safety) and safety work to which the SMS belongs. They suggest looking at safety work through institutional work lenses, beyond supporting the safety of work, thus the enhancement of safety performance. They suggest four types of safety work: social safety (conceptual work aiming at maintaining safety as a value), demonstrated safety (structural work aiming at showing that the organization fulfills its safety obligations), administrative safety (structural work oriented inward to influence operational work), and physical safety (directly transforming the work environment). "Managers and workers perform both 'Safety Work' and 'Operational Work'—they co-create the institution that governs their day-to-day lives" (Rae & Provan, 2019, p. 121). The authors claim that considering safety work as institutional work is necessary "to acknowledge the social complexity of modern organisations" (Rae & Provan, 2019, p. 126). Pushing the idea that the SMS serves another purpose than enhancing safety performance (namely, an institutional one) and encourages one to widen the lenses through which the SMS can be looked at, even further than safety itself, to explore whether other purposes can be identified. Investigating the context in which the SMS emerged and spread might bring yet another version of the SMS and perspective on its *raison d'être*.

NOTES

1. Field note made during a meeting discussion on the SMS in the framework of the Future Sky Safety European project.
2. Field note made during a meeting discussion on the SMS in the framework of the Future Sky Safety European project.
3. UK CAA webpage on SMS strategy (www.caa.co.uk/Safety-initiatives-and-resources/Working-with-industry/Safety-management-systems/Safety-management-systems-strategy).
4. Field note made during a meeting discussion on the SMS in the framework of the Future Sky Safety European project.
5. For example, the Tōhoku earthquake in March 2011 and the resulting tsunami that are environmental hazards led to the Fukushima Daiichi nuclear power plant explosion that is safety consequences.
6. Conversely, a safety event like the core melt and explosion of Fukushima Daiichi nuclear power plants in March 2011 led to the release of radioactive material in the environment—that is, environmental consequences.
7. The "global risk analysis" method suggests to start from a systematic exploration of 26 generic hazards related to, respectively, (1) the environment of the organization (politics, environment, security, image, customers); (2) the organization's governance (e.g., management, finance, commercial, strategic); (3) the organization's technical means (infrastructure and facilities, tools and equipment, information system); and (4) the operations of the organization (e.g., operations, human performance, products). From the identification of specific hazards based on this list, the method inductively develops possible accident scenarios that in turn allow for assessing and prioritizing, thus managing the risks.
8. This translates as "The field of management is not limited to that of administration."
9. See www.caa.co.uk/Safety-initiatives-and-resources/Working-with-industry/Bowtie/About-Bowtie/Introduction-to-bowtie.
10. Using the UK CAA templates is not ideal here either. However, according to my observations in several aviation organizations, the bow-tie diagrams developed internally derive from the same approach and are even sometimes a mere cut-and-paste of the bow-tie templates provided by the CAA. However, for confidentiality reasons, they cannot be reproduced in this document.
11. See www.icao.int/safety/SafetyManagement/SMI/Documents/BowTieXP%20Methodology%20Manual%20v15.pdf.

REFERENCES

Acfield, A. P., & Weaver, R. A. (2012). *Integrating safety management through the Bowtie concept: A move away from the safety case focus*. Proceeding of the Australian System Safety Conference (ASSC 2012) (Vol. 145).

Airbus. (2020). *A statistical analysis of commercial aviation accidents 1958–2019*. Airbus.

Alizadeh, S. S., & Moshashaei, P. (2015). The Bowtie method in safety management system: A literature review. *Scientific Journal of Review*, *4*(9), 133–138.

Almklov, P. G. (2018). Situated practice and safety as objects of management. In: *Beyond safety training* (pp. 59–72). Springer.

Almklov, P. G., Rosness, R., & Størkersen, K. (2014). When safety science meets the practitioners: Does safety science contribute to marginalization of practical knowledge? *Safety Science, 67*, 25–36.

Amalberti, R. (2017). *La sécurité industrielle est-elle un art du compromis ? Audit, risques & contrôle, 12*.

Andéol-Aussage, B., Drais, E., & Montagnon, C. (2018). *Le management de la santé et sécurité au travail (S&ST): Particularismes et évolutions*. Hygiène et sécurité du travail—n° 253—décembre 2018.

Aven, T. (2014). *Risk, surprises and black swans: Fundamental ideas and concepts in risk assessment and risk management*. Routledge.

Baram, M., & Lindoe, P. (2018). Risk communication between companies and local stakeholders for improving accident prevention and emergency response. In: *Risk communication for the future* (pp. 61–77). Springer.

Bastide, L. (2018). Crisis communication during the Ebola outbreak in West Africa: The paradoxes of decontextualized contextualization. In: *Risk communication for the future* (pp. 95–108). Springer.

Bieder, C. (2017). Conclusion. In: *The illusion of risk control* (pp. 107–112). Springer.

Bieder, C., & Bourrier, M. (2013). *Trapping safety into rules: How desirable or avoidable is proceduralization?* Ashgate-CRC Press.

Bieder, C., & Callari, T. C. (2020). Individual and environmental dimensions influencing the middle managers' contribution to safety: The emergence of a "safety-related universe." *Safety Science, 132*, 104946.

Bourrier, M. (2005). The contribution of organizational design to safety. *European Management Journal, 23*(1), 98–104.

Bourrier, M. (2017). Organisations et activités à risque: le grand découplage. In: Barbier, J.-M. & Durand, M. (eds.), *Analyse des activités humaines. Perspective encyclopédique* (pp. 743–774). Presses Universitaires de France.

CAA. (2014). *Safety Management Systems (SMS) guidance for organisations*. Document CAP 795. CAA. https://publicapps.caa.co.uk/modalapplication.aspx?appid=11&mode=detail&id=6616

Cacciabue, P. C., Cassani, M., Licata, V., Oddone, I., & Ottomaniello, A. (2015). A practical approach to assess risk in aviation domains for safety management systems. *Cognition, Technology & Work, 17*(2), 249–267.

Casner, S. M., Geven, R. W., Recker, M. P., & Schooler, J. W. (2014). The retention of manual flying skills in the automated cockpit. *Human Factors, 56*(8), 1506–1516.

Cochoy, F., & De Terssac, G. (2009). Les enjeux organisationnels de la qualité: une mise en perspective. In: *Sciences de la Société* (pp. 3–18). Presses Universitaires du Midi.

Dekker, S. W. (2014). The bureaucratization of safety. *Safety Science, 70*, 348–357.

Dekker, S. W. (2019). *Foundations of safety science: A century of understanding accidents and disasters*. Routledge.

Desroches, A., Aguini, N., Dadoun, M., & Delmotte, S. (2016). *Analyse globale des risques: principes et pratiques*. Lavoisier.

Desroches, A., Baudrin, D., & Dadoun, M. (2009). *L'analyse préliminaire des risques: principes et pratiques*. Lavoisier.

Desroches, A., Leroy, A., & Vallée, F. (2003). *La gestion des risques*. Lavoisier.

Dupré, M., Etienne, J., & Le Coze, J. C. (2014). *The regulator-regulatee interaction: Insights taken from a risk-laden business firm*. 2. Annual Cambridge Conference on Regulation, Inspection & Improvement: The End of Zero Risk Regulation: Risk Toleration in Regulatory Practice, Cambridge, United Kingdom.

Dupuy, J. P., & Grinbaum, A. (2005). Living with uncertainty: From the precautionary principle to the methodology of ongoing normative assessment. *Comptes Rendus Geoscience, 337*(4), 457–474.

Fucks, I., & Dien, Y. (2013). "No rule, no use"? The effects of over-proceduralization. In: *Trapping safety into rules how desirable or avoidable is proceduralization* (pp. 27–39). CRC Press.

Grote, G. (2012). Safety management in different high-risk domains: All the same? *Safety Science, 50*(10), 1983–1992.

Gunningham, N., Kagan, R. A., & Thornton, D. (2004). Social license and environmental protection: Why businesses go beyond compliance. *Law & Social Inquiry, 29*(2), 307–341.

Hacking, I. (2001). *Leçon inaugurale—Chaire de philosophie et histoire des concepts scientifiques*. Collège de France, 11 janvier.

Heimann, L. (1997). *Acceptable risks, politics, policy and risky technologies*. The University of Michigan.

Heimann, L. (2005). Repeated failures in the management of high risk technologies. *European Management Journal, 23*(1), 105–117.

Hollnagel, E. (2014). *Safety-I and safety-II*. Ashgate.

Hovden, J. (1998). The ambiguity of contents and results in the Norwegian internal control of safety, health and environment reform. *Reliability Engineering & System Safety, 60*(2), 133–141.

Hovden, J., & Tinmannsvik, R. K. (1990). Internal control: A strategy for occupational safety and health: Experiences from Norway. *Journal of Occupational Accidents, 12*(1–3), 21–30.

Hudson, P. T. W. (2001). Safety management and safety culture: The long, hard and winding road. In: *Occupational health and safety management systems* (pp. 3–32). Crowne Content.

ICAO. (2009). *Safety management manual (SMM)* (2nd ed.). ICAO.

ICAO. (2013). *Annex 19: Safety management* (1st ed.). ICAO.

ICAO. (2016). *Annex 19: Safety management* (2nd ed.). ICAO.

ICAO. (2018). *Safety management manual (SMM)* (4th ed.). Doc 9859. ICAO.

ISO. (2009). *Guide 73: Risk management: Vocabulary*. International Organization for Standardization.

Jasanoff, S., (2003). Technologies of humility: Citizen participation in governing science. *Minerva, 41*, 223–244. Kluwer Academic Publishers.

Kaspers, S., Karanikas, N., Piric, S., van Aalst, R., de Boer, R. J., & Roelen, A. (2017). Measuring safety in aviation: Empirical results about the relation between safety outcomes and safety management system processes, operational activities and demographic data. In: *PESARO 2017: The Seventh International Conference on Performance, Safety and Robustness in Complex Systems and Applications* (pp. 9–16). IARIA.

Kjellén, U. (2009). The safety measurement problem revisited. *Safety Science, 47*, 486–489.

La Porte, T. R. (2020). Doing safety . . . and then security: Mixing operational challenges: Preparing to be surprised. In: *The coupling of safety and security* (pp. 75–85). Springer.

Le Coze, J.-C. (2005). Are organisations too complex to be integrated in technical risk assessment and current safety auditing? *Safety Science, 43*(8), 613–638.

Le Coze, J.-C. (2011). *De l'investigation d'accident à l'évaluation de la sécurité industrielle: proposition d'un cadre interdisciplinaire (concepts, méthode, modèle)*. Gestion et management. Ecole Nationale Supérieure des Mines de Paris.

Leveson, N., Dulac, N., Marais, K., & Carroll, J. (2009). Moving beyond normal accidents and high reliability organizations: A systems approach to safety in complex systems. *Organization Studies, 30*(2–3), 227–249.

Li, Y., & Guldenmund, F. W. (2018). Safety management systems: A broad overview of the literature. *Safety Science, 103*(2018), 94–123.

Macrae, C. (2009). Making risks visible: Identifying and interpreting threats to airline flight safety. *Journal of Occupational and Organizational Psychology, 82*(2), 273–293.

Macrae, C. (2019). Moments of resilience: Time, space and the organisation of safety in complex sociotechnical systems. In: Wiig, S. & Fahlbruch, B. (eds.), *Exploring resilience: Briefs in applied sciences and technology*. Springer.

Madsen, P. M. (2013). Perils and profits: A reexamination of the link between profitability and safety in US aviation. *Journal of Management, 39*(3), 763–791.

March, J. G., & Simon, H. A. (1958). *Organizations*. Wiley.

Maurino, D. (2017). *Why SMS: An introduction and overview of safety management systems*. OECD. www.itf-oecd.org/sites/default/files/why-sms.pdf

Meadows, D. H. (2008). *Thinking in systems: A primer*. Chelsea Green Publishing.

Millstone, E., & Van Zwanenberg, P. (2000). A crisis of trust: for science, scientists or for institutions? *Nature Medicine, 6*(12), 1307.

NTSB. (2009). *Accident report, NTSB/AAR-10/01, PB2010–910401*. www.ntsb.gov/investigations/AccidentReports/Reports/AAR1001.pdf

Pelegrin, C. (2013). The never-ending story of proceduralization in aviation. In: *Trapping safety into rules: How desirable or avoidable is proceduralization* (pp. 13–25). CRC Press.

Pettersen Gould, K. (2020). *Organizational risk: "Muddling through" 40 years of research: Risk analysis*. https://doi.org/10.1111/risa.13460

Popov, G., Lyon, B. K., & Hollcroft, B. (2016). *Risk assessment: A practical guide to assessing operational risks*. John Wiley & Sons.

Rae, A., & Provan, D. (2019). Safety work versus the safety of work. *Safety Science, 111*, 119–127.

Rasmussen, J. (1997). Risk management in a dynamic society: A modelling problem. *Safety Science, 27*(2), 183–213.

Reason, J. (1997). *Managing the risks of organizational accidents*. Ashgate.

Reason, J. (1999). Are we casting the net too widely in our search for the factors contributing to errors and accidents. In: *Nuclear safety: An ergonomics perspective* (pp. 199–205). CRC Press.

Reiman, T., & Pietikäinen, E. (2012). Leading indicators of system safety: Monitoring and driving the organizational safety potential. *Safety Science, 50*(10), 1993–2000.

Sankar, V. K., & Siddiqui, N. (2016), Application of bow-tie analysis in risk management. *International Journal for Scientific Research & Development, 4*(9), 212–214.

Schulman, P. R. (2020). Integrating organizational and management variables in the analysis of safety and risk. In: *Human and organisational factors* (pp. 71–81). Springer.

Schulman, P. R., Roe, E., Eeten, M. V., & Bruijne, M. D. (2004). High reliability and the management of critical infrastructures. *Journal of Contingencies and Crisis Management, 12*(1), 14–28.

Simon, H. A. (1957). *Administrative behavior* (2nd ed.). Macmillan.

Stolzer, A. J., & Goglia, J. J. (2015). *Safety management systems in aviation* (2nd ed.). Ashgate.

Taleb, N. N. (2007). *The black swan: The impact of the highly improbable* (Vol. 2). Random House.

Thomas, M. J. (2012). *A systematic review of the effectiveness of safety management systems* (No. AR-2011–148). Australian Transport Safety Bureau.

Turner, B. A., Pidgeon, N., Blockley, D., & Toft, B. (1989, November). Safety culture: Its importance in future risk management. In: *Position paper for the second World Bank workshop on safety control and risk management, Karlstad, Sweden* (pp. 6–9).

Voss, B. (2012, May). SMS reconsidered. *Aerosafety World*, p. 1.

Wachter, J. K. (2011). Ethics: The absurd yet preferred approach to safety management. *Professional Safety, 56*(6), 50.

Weick, K., & Sutcliffe K. (2015). Managing the unexpected: Sustained performance in a complex world (3rd ed.). Wiley.

Wreathall, J. (2009). Leading? Lagging? Whatever! *Safety Science, 47*, 493–494.

5 Why Did the SMS Emerge and Spread?

Considering the "disillusion" that SMS brings as an allegedly revolutionary safety approach, its success as a common and widely accepted safety approach must be based on other forms of explanation. Since the notion of SMS did not emerge in aviation and appeared long before it was adopted in aviation, it is interesting to go back to the origins of the SMS to try and understand why and how it emerged and spread to a point where it was considered, and still is, to a certain extent a promising way forward in safety. This chapter explores the genesis of the SMS. First, the changes in high-risk industries posing new safety challenges are reviewed. Whether technological, organizational, or more related to the economic context, the 1980s were rich in novelties. Besides these objective evolutions, we examine the motivations of the various safety stakeholders to move toward a new safety approach. After investigating the push for a new safety approach, why all the actors converged toward an approach like the SMS remains to understand. Therefore, we analyze the global context and trends of that time that could have contributed to this convergence. In addition, we investigate the way SMS-related ideas traveled and spread. Considering the delay between the emergence of the SMS and its adoption in aviation, we perform a specific analysis to understand the reasons underlying this late adoption.

The main source of evidence underlying this chapter, beyond existing literature, is a series of interviews with individuals who played a personal part in safety at the time the SMS started to be discussed and were observers or actors of the emergence and spreading of the SMS.

This study allows for unveiling the many purposes that SMS fulfills, especially the unspoken ones, beyond the declared safety-related promises, that contributed to propelling the SMS as a key turning point in safety management. The SMS appears as the minimum consensual satisficing dispositive.[1] It allows for the meeting of all stakeholders' objectives, from safety enhancement to liability, transparency, loss control, or overall performance improvement. At the same time, the study points out that the emergence of the SMS is situated in time but also in a broader industrial, economic, political, and social context (Bieder, 2021).

5.1 NEW SAFETY CHALLENGES

The limitations of the historical approach to safety management, based on compliance with rules, norms, procedures, and linear analysis started to be pointed out in the 1980s as a result of some significant trends leading to new natures of risks (Perrow, 1984; Rasmussen, 1997). The changes highlighted in the literature are of a diverse nature, depending on the authors and their perspective, but all ultimately call for a new approach to safety management.

DOI: 10.1201/9781003307167-5

5.1.1 Technological Evolution

Among the highlighted changes, the advent of new information technology plays an important role in several respects and creates significant vulnerabilities in many areas (Rochlin, 1997). It increases the exchange of information and communication, thus the complexity of systems and coupling between systems (Rasmussen, 1997; Leveson, 2004), one of the main reasons calling for a systemic approach to safety (Rasmussen, 1997; Leveson, 2004; Perrow, 1984; Hollnagel, 2008). It also introduces new modes of failure, new hazards, and new risks, and thus ultimately unknowns and even unknown unknowns (Leveson, 2004; Hollnagel, 2008). The increasing complexity of systems also leads to considering that accidents may result from non-linear interactions between normal variability of performance as much as from consequences of failures or malfunctions (Perrow, 1984; Hollnagel, 2008).

Another aspect of the fast pace of technological change is the increasing potential for wide-scale accident and harm, not only through immediate consequences but also through long-term consequences (Lagadec, 1981; Rasmussen, 1997).

5.1.2 Fragmented and Extended Enterprises

In parallel to this technological change, the literature refers to other evolutions likely to affect safety, such as the changing nature of enterprises. The tendency for big companies over the past decades is toward more and more fragmentation and the emergence of an extended enterprise, whereby focusing on a particular organization is debatable and the very boundaries of an organization are blurred (joint ventures, branches of the same group). Lehtinen and Ahola (2010) date this evolution back to the end of the 20th century, driven by the internationalization and opening of new markets that took place in East Europe and Asia during the 1980s and 1990s, forcing organizations to seek collaboration with local partners and leverage resources through sharing them. The development of information technology was also a key enabler for inter-organizational collaboration, allowing for the exchange of huge amounts of information in real time.

Dien and Dechy (2013), while analyzing this increasing fragmentation of organizations, emphasize that it translates into an excess of separation, of fragmentation into smaller organizations to make them more operational and efficient but with increasingly complex relationships at the interfaces. To add to this already challenging situation, they mention that fragmentation is dynamic and changes according to the industrial policy at a given point in time. With their definition of fragmentation, several distinct entities linked by contract or formal links collaborating toward a common global objective, a number of aviation stakeholders can be considered fragmented organizations, with tier 1, tier 2, and tier 3 suppliers, joint ventures, and subcontractors.

From a safety viewpoint, fragmentation of organizations was identified as a challenge in many respects and adds complexity to an already complex issue (Dien & Dechy, 2013; Milch & Laumann, 2016). Among the induced challenges identified in several sources (Dien & Dechy, 2013, pp. 10–11; Milch & Laumann, 2016, pp. 13–16) are the deterioration of the sense of responsibility for safety, coordination and communication problems at the interface, and the tendency to reinforce the formal work and adopt a work-by-the-book approach to compensate for the lack of knowledge of real hazards and safety risks on the job and manage the relationship between actors. Other difficulties were highlighted, such as the development of a shared safety culture and shared safety measures or, even more generally, shared goals, for different organizations have different goals. The emergence of a blame culture in which organizations blame each other if something goes wrong was also noticed as an inter-organizational safety challenge, leading to a lack of reporting and sharing of experience, and even employees not daring to ask critical questions in tricky situations in such defiance and distrust atmosphere. For these various reasons, fragmentation of organizations has been pointed out as a contributor to a number of accidents (Challenger space shuttle accident in 1986; in-flight collision over Überlingen in 2002) and, more generally, as inducing safety challenges that are to be taken into account in approaches aiming at managing safety.

5.1.3 Interconnected Infrastructures and Organizations

Interconnected infrastructures and organizations largely developed in the 1980s. It was the case, for example, of electrical power distribution systems, water distribution systems, or transport systems. The interconnection of infrastructures and organizations to deliver a service (including critical ones such as the ones mentioned earlier) induces new sources of complexity, thus of criticality, as argued by Bouwmans et al. (2006) and Weijnen and Bouwmans (2006): (1) a physical complexity with technical systems and physical infrastructures being connected to one another and emerging behaviors being observed from these interrelations; (2) a social complexity with organizations having their own interests and objectives being intertwined with the shared objectives of the interconnected organizations. Last, strong interdependencies of infrastructures that cannot function without the others. Along Perrow's lines (1984), for such a complex and tightly coupled system, accidents should be "normal," even more so considering that many of these organizations have competing if not conflicting goals, and there is no ongoing command and control despite the presence of rapidly changing task environments with highly consequential hazards (Schulman et al., 2004).

The air transport system, with its multiple and distinct stakeholders/organizations simultaneously involved in and interrelated to make air operations possible, safe, and efficient, can definitely be considered an interconnected infrastructure (Maurino, 2017; ICAO, 2016).

5.1.4 An Increasing Economic Pressure

Beyond the technology and the organizations and enterprises themselves, the economic context in which companies evolve is changing as well, becoming aggressive and very competitive, and ultimately focusing "the incentives of decision makers on short term financial and survival criteria rather than long term criteria concerning welfare, safety, and environmental impact." This leads companies to operate at the edge and makes a safety management approach based on chasing deviations irrelevant (Rasmussen, 1997, p. 186).

Safety and productivity are commonly simply opposed and considered conflicting stakes for organizations.[2] Yet the relationship between the two is more complex than that. They can also, under some conditions, be perceived as constituent of one another. Madsen (2013) shows in a study of US airlines over a period from 1990 to 2007 that "for firms performing below their profitability aspiration, a positive relationship exists between profitability and accident rate" such that when profitability just misses meeting aspirations, decision-makers may see in the reduction of safety expenditures a way to meet the profitability expenditure. Conversely, for firms performing above their profitability aspirations, there is a negative relationship between profitability and the accident rate. Indeed, "for decision makers in organizations performing far above their profitability aspirations, the difference between minimal and generous safety investments should not alter the prospect of meeting future goals, but a major accident could easily damage future profitability" (Madsen, 2013, p. 772). In brief, "accidents and incidents are more likely to be experienced by organizations performing near their profitability targets." In other words, the tendency of organizational decision-makers to take safety-related risks is related to profitability aspirations more than to absolute levels of profitability (Madsen, 2013, p. 785). If these results could be seen as a plea against current management systems and associated indicators that lead certain managers and senior managers to manage indicators and targets more than the company itself, it reflects the close interrelation between safety and other stakes within organizations.

5.2 THE INTELLECTUAL CONTEXT AROUND SAFETY: DIVERSE SCHOOLS OF THOUGHT, EVOLVING CONCEPTUAL FRAMES

The SMS is not a revolutionary idea that spontaneously appeared. As mentioned earlier, the first discussions around SMS date from the 1970s, even if the approach spread widely in the 1980s and 1990s, especially in the process industries (e.g., chemistry; oil and gas), and reached aviation almost two decades later. The intellectual context around safety management was dynamic, with a variety of scientific developments running at the same time. In order to understand the context of ideas on safety in which the SMS emerged, a historical overview of the various approaches and theories related to safety management provides useful insights into the safety thinking landscape and its evolution throughout this period. Without reviewing the whole history of safety science, three major scientific strands particularly stand out when it comes to managing safety and understanding where the SMS took its inspiration from and what it disregarded. Hardly ever exchanging with one another, these parallel strands formed the intellectual context around safety in the 1980s/1990s.[3]

In 1985, Rasmussen already distinguished between two worlds of research that hardly ever managed to work together if only to talk (Rasmussen, 1985): on the one hand, engineers coming from the technological side of safety, adopting a systems' engineering and risk analysis angle to safety, and on the other hand, researchers in human factors, initially focusing on the individual behavior and later on the managerial contribution to accidents. However, considering the developments in social sciences in the late 1980s, a third scientific strand can be added, with social scientists focusing on the organizational and social contributions to safety. The various schools of thought are presented successively in the following sections.

5.2.1 The Risk Management Perspective on Safety

Risk management was initially developed in the mid-20th century to control the risks associated with technological systems (especially in the military and the space industry). In that respect, it aimed to support managers' decision-making process by clearly identifying the risks that are unacceptable as such and proposing risk mitigation measures and a way to control that they are implemented and efficient. Historically, risk analysis methods were used in the design phase initially to improve high-risk technological systems' design and intrinsic safety and, more generally, their dependability (i.e., reliability, availability, maintainability, and safety). Beyond design, they were also used as a support to internal safety risk management.

The extension of their scope to the "human component" dates back to the 1960s with the analysis and quantification of "human error" as the human contribution to man-machine systems failure. The first widespread approach along these lines—that is, the HRA method, developed in the US at Sandia National Laboratory—was a *technique for human error prediction* (THERP) (Swain, 1964).

Until the 1990s, most risk analyses that integrated the human component did it by focusing on first-line operators and their errors as any other technical failures (Everdij & Blom, 2016; Swain, 1990). These approaches were mainly aimed at supporting the design of or certifying new technological systems. More generally, they were part of an overall risk management approach focused on safety risks (Desroches et al., 2003).

Critically, the scope of risk management at the foundation of these methods is often forgotten or at least remains implicit, whereas it is enlightening with regards to some recurring debates. Indeed, risk management focuses on the known and partly on the knowable spaces, recognizing the existence of an unknowable space that falls outside of its scope; this space being unknown, it cannot be controlled (Desroches et al., 2003). In the 1990s, some organizational and managerial aspects were integrated into risk analysis. These ranged from the addition of some organizational factors as hazards (Mohaghegh & Mosleh, 2009; Desroches, 2009; Everdij & Blom, 2016) to implementing systems engineering methods to human and organizational aspects (e.g., HAZOP) or developing ad hoc methods to assess the reliability of human and organizational contribution to operations (e.g., advanced HRA methods such as ATHEANA and MERMOS). Interestingly, the later methods, although ultimately quantitative, mainly involved a qualitative analysis.

5.2.2 The Human Factors and Managerial/Organizational Perspective on Safety

Academics coming from the individual behavior scope, such as psychologists, especially Andrew Hale, James Reason, and Bernhard Wilpert, or with a mixed profile, such as Jens Rasmussen, started focusing in the 1970s and 1980s on the understanding of individual behavior and human error and its contribution to accidents (Rasmussen, 1982, 1983; Reason, 1988; Hale & Glendon, 1987). They had some interactions with academics in systems engineering and risk analysis, willing to refine their modeling and quantification of human error (Rasmussen, 1985). Initially focusing on individuals, they broadened their scope in the late 1980s to explain accidents of socio-technical systems and introduced a more managerial-like approach to academic research. The most popular illustration of this evolution is the Swiss cheese model introduced by Reason (1990), which highlights the key role played by management in ensuring working conditions that influence unsafe acts. The tenuous link between these academics coming from human error (and thus known for some of them from the HRA and systems engineering community) and the ones in systems safety also contributed to the attempt to introduce organizational aspects into risk analysis. It initially translated into the introduction in risk analysis of organizational errors as an extension of the idea of human error.[4]

Rasmussen, also inspired by complexity science, suggested in the late 1980s to view safety as a control problem, which led him to propose in 1997 a hierarchical control model of the socio-technical system involved in risk management reaching even beyond the managerial level to encompass as well the company, regulators, and the government (Rasmussen, 1989, 1997). The modeling of safety as a control problem creates a natural link with systems engineering approaches and the integration of organizational aspects into risk analysis.

However, in the organizational evolution of the human factors approaches, organizations are not necessarily viewed and understood as they are by social scientists, as the latter's developments in the late 1980s and 1990s illustrate.

5.2.3 The Organizational and Social Perspective on Safety

The historical evolution of social science approaches to safety was extensively described and discussed by several authors (Le Coze, 2012; Bourrier, 2017). The purpose here is not to repeat this discussion but rather to highlight the elements allowing for putting these approaches into perspective with the ones deriving from the risk management and human factor's angles to safety.

Whether because the nature of accidents evolved, making approaches focusing on individuals too limited, or because the perspective on the analysis and understanding of accidents changed, the 1980s were a decade where a number of major accidents occurred (e.g., the Challenger disaster and the Chernobyl disaster, both in 1986) and attracted the interest of social scientists, especially sociologists. Their purpose was to describe, understand, and explain how collectives and organizations contribute to accidents or to safety.

These two different angles led to two types of contributions. A first contribution, starting from accident, led to an analysis of how high-risk organizations drifted and/or failed. For example, Barry Turner's theory identifying a long period of incubation before the occurrence of accidents where the organization remains blind to important issues was published in 1978 in the book *Man-made Disasters* (Turner, 1978). Similarly, organizational mechanisms whereby deviance became normalized were described by Diane Vaughan in 1996.

The second type of contribution, starting from a positive angle, analyzed how organizations successfully managed to sustain a high-reliability performance like, for example, La Porte, Schulman, and the whole Berkeley school of thought on high-reliability organizations (HROs). This scientific development started in the late 1980s with extensive field work in three different high-risk organizations apparently performing in a reliable way, including safety-wise. Beyond the role of the organization in the way operations are handled, and thus their safety managed, these approaches also point out the role of external actors and aspects that also interplay with operations. Regulatory and oversight authorities are an obvious example, but the mutual influence reaches beyond these institutions. A number of watching groups, both formal and informal, contribute to maintaining a high interest in highly reliable operations (La Porte, 1996). More broadly, Jasanoff (2003) underlines the value of the public sphere inputs even more so in areas like health, safety, and environment, where there is a high degree of uncertainty. If uncertainty is not exclusively due to a lack of knowledge, traditional risk governance based on risk assessment relies on a limited and simplified knowledge. Limited because relying on a limited number of scientific experts but also simplified because relying on generic models (Jasanoff, 2003).

It is important to underline that while the risk analysis approaches mentioned earlier were referring to the "offline" management of safety, or kind of back-office activities, a number of social science approaches focused on the "online" management of safety—that is, the role of operational workers and managers in real-time production, like the HRO theory.[5] Moreover, the interdependence between production and safety and some kind of uncertainty were founding elements of the HRO theory. Indeed, it focuses on the twofold operational challenges that is to "manage complex, hazardous, demanding technologies while avoiding major failures; and, at the same time, to maintain the capacity for meeting intermittent, somewhat unpredictable, periods of high peak demand and production" (La Porte, 1996, p. 60). An organization is viewed not as a rigid structure but as a living being (Bourrier, 2017, 2005; Weick & Sutcliffe, 2015; Amalberti, 1996) where adaptation and flexibility are encouraged. Hierarchy is not the only way the organization functions, especially when expertise is needed. In other words, HROs tend to look at organizations in a systemic way, acknowledging tensions and contradictions, as well as uncertainty, and identifying circular patterns of influence. Interestingly, the HRO theory does not focus on safety but on how the tension between production and safety holds, including in unexpected situations.[6] However, the aim was initially to describe the functioning of these "apparently" highly reliable organizations, even though it was then considered and used by some consultants and scholars as an explanatory and even a normative model. More generally, organizational and social approaches were rather aimed at describing and explaining how safety was "managed" than being prescriptive and, for example, directly supporting design or management as such.

Transitioning from description to prescription is not immediate if at all possible. As stated by Pidgeon (2010, pp. 212–213),

> By 1990, it was clear that the cutting-edge intellectual focus was less on analysing how past accidents had occurred (important though that issue still was) and more towards the question of how safe organisations might be encouraged or even designed. . . . But it was also clear even then that understanding how vulnerability to failures and accidents arises does not automatically confer predictive knowledge to prevent future catastrophes. In effect, many of us began to ask whether a theory of vulnerability to error and failures could be transposed into one of practical resilience.

5.3 A VARIETY OF MOTIVATIONS TO MOVE TOWARD A NEW SAFETY MANAGEMENT APPROACH

The emergence of new technological, organizational, or economic safety hazards and risks calls for a new safety management approach. Despite the availability of a wealth of concepts and theories on accidents and safety, industrial actors do not necessarily interact with academics. Yet other aspects are at play in the emergence of safety management approaches. This section explores the motivations of the various safety stakeholders to change their approach to enhance safety around the 1980s.

5.3.1 Insurance Companies: Better Calibrating Premiums

Insurance companies are among the stakeholders that have always been interested in reducing the damage caused by accidents. Beyond any ethical considerations, one of the stakes for these companies is the damage related to accidents. Interestingly, the function of risk manager was first introduced by Gallagher, from an insurance company, in the *Harvard Business Review* (Gallagher, 1956). Insurance companies historically focused on workplace accidents and then extended their scope to industrial accidents. This permanent concern with safety management appeared significantly earlier than the 1970s or 1980s. One of the most famous legacies of the effort of insurance companies to reduce the number and severity of accidents is the accident pyramid and the accident causation model (domino model), published by Herbert Heinrich (1931). Interestingly, one of the dominos already looked beyond the worker by referring to his/her "social environment" as a cause for unsafe acts. Another main contribution to the understanding and modeling of accidents, and as a follow-up to safety management, was put forward in the 1970s by Frank Bird Jr. when he was recruited by an insurance company after the first phase of his career as an engineer and then as a safety manager in the steel industry in the US. In some US states, at that time, there was no public insurance. Therefore, companies needed to subscribe to private insurance. Bird took the modeling effort started by Heinrich one step further and developed a tool that could support insurance companies in calculating a company's insurance premium based on what was put in place by these companies to manage safety (Bird et al., 1974). "What he had developed served as a structure to determine the level of risk of a company, thus its insurance premium. Frank Bird tried to put criteria to determine this premium" (consultant/industry 2, diverse industries, December 21, 2018). These criteria included a number of organizational and managerial arrangements that were considered by Bird (through his analysis of thousands of reported accidents) as good safety management practices.[7]

This approach, formerly aimed at assessing companies' safety management capabilities and practices, also played a role in the motivation of some industrial companies to move toward a new approach since it was turned in the late 1970s into a tool to support companies themselves to better manage their safety, namely, the International Safety Rating System (ISRS),[8] voluntarily adopted by a number of high-risk industries.

5.3.2 Industrials: Trauma, Ethics, and Performance

Just like for insurance companies, accidents are part of the motivation for industrials to move toward new safety management approaches. As mentioned by one of the former employees of the consulting company set up by Bird in the late 1970s to disseminate his ISRS tool,

> Their [the companies we were working for] initial motivation was often triggered by a trauma, either internal or close to the company or experienced by one of the organization's leader in his previous company. The change to adopt the ISRS was not spontaneous, it was rather an electroshock.
>
> **(Consultant/industry 2, diverse industries, December 21, 2018)**

However, other motivations were mentioned by the interviewees regarding the decision to develop or adopt different approaches to further enhance safety in the 1990s, including ethical reasons, beyond the reduction of the costs associated with accidents.

As stated by one of the interviewees,

> Management systems were already discussed at that time internally. Chemistry was a leader in terms of safety and safety management. HSE as well as social aspects were explicitly part of the values promoted by the company.[9] It was agreed at the top level of the company that people were working at [name of the company] to make a living, not to lose their lives. It wasn't just words.
>
> **(Industrial 2, chemistry, January 25, 2019)**

In the 1990s, other industries in different countries as well funded their own research and development to further enhance their approaches to managing safety. It was the case of Shell in the oil and gas industry that wanted to develop its own methods to identify organizational contributors to safety and performance more generally. Enhancing safety management was seen as a way to improve their overall performance by some companies and formed part of their motivations to move toward a new approach. In the case of the adoption of the ISRS, "safety was just a pretext. It consisted in expecting from each manager to bring its contribution to the good functioning of the company and to take his share of responsibility" (consultant/industry 2, diverse industries, December 21, 2018). More generally, there was an overall organizational challenge. As underlined by one of the interviewees from the oil and gas industry,

> The system is complex. In a refinery, there are around 40 different units, different temperatures and pressures over a wide range, a variety of materials, around 10 000 instruments. The operators don't have a high level of education (conversely to pilots). For example, in [one of our sites] back in the seventies, we recruited people with all types of background (e.g. butchers, bakers etc.). Industry trained them to become an operator. Hence the main question was not so much about behavior but more about how to get this organized.
>
> **(Mixed experience 4, oil and gas, September 17, 2020)**

In the nuclear industry, where the main approach to safety was certification, EDF (Électricité de France) in France and the NRC (Nuclear Regulatory Commission) in the US decided in the early 1990s to develop new HRA methods to integrate "online" safety management aspects and operating experience. Performing these more sophisticated analyses would not only respond to the mandatory certification requirements but also, beyond the certification exercise, allow the drawing of useful insights to enhance safety (Bieder et al., 1998; Cooper et al., 1996). In France, the trigger was a change in technology with the introduction of computerized control rooms and the need to certify these new N4 reactor types with these novel control rooms that were being built in Chooz et Civaux (respectively northeast and center-west of France). These HRA developments were focused on enhancing safety by the design of a socio-technical system rather than by supporting managers in addressing safety.

5.3.3 Regulators: Overcoming the Pitfalls of Command and Control

Different motivations were also identified on the part of regulatory agencies, which are not all similar across countries and hazardous industries or did not appear at the same time. Coming from different regulatory regimes, several countries have played an influential role in the development of a new approach to safety management, such as Norway (especially the oil and gas industry there), the Netherlands, and the UK.

The reconsideration of the "command and control" model of regulation dates from the end of the 1960s (Aalders & Wilthagen, 1997). It translated into some influential reports, such as that by the Robens Committee in the UK that was published in 1972, and criticized the traditional regulation model, mainly a top-down one, or the introduction of the "humanization of work" concept in the Netherlands inspired by the Norwegian Work Environment Act published in 1977 (Aalders & Wilthagen, 1997).

The Norwegian Case

The Norwegian Petroleum Directorate (NPD), established in the 1970s to regulate and oversee the emerging oil and gas activity, soon initiated reflections on the regulatory regime to be adopted in their particular context, where (Lindøe & Olsen, 2009)

- the Norwegian Work Environment Act of 1977, "the world's possibly most stringent labour legislation" (ibid., p. 432), was in place;

- the offshore industry was densely unionized "with extensive collective bargaining rights and a comprehensive network of safety representatives" (ibid., p. 432); and
- two major accidents occurred (the blow-out on the Bravo platform in 1977 and the capsizing of the Alexander Kielland platform in 1980), showing the limitations of the regulatory regime in place.

The NPD pushed for moving from a reward and punishment to a mutual understanding and cooperation approach to safety regulation (Lindøe & Olsen, 2009; Hovden & Tinmannsvik, 1990).

This reflection initially gave rise to an internal control approach that was then extended to all industries to overcome a double bind resources issue.[10] The first issue is a qualitative one related to the increase in complexity and automation in industry, making it hard for a regulatory agency to keep up with. The second issue is a quantitative one since resources were becoming too limited to continue traditional monitoring (Hovden & Tinmannsvik, 1990). The stake was then to perform lighter reviews that would make the authorities confident that industries were managing safety correctly, thus, moving from traditional regulation to the monitoring by authorities of industries' self-regulation (Hale & Hovden, 1998). As such, internal control leads to a new balance between resources and responsibility, more specifically to an evolution from detailed prescriptive laws and regulations specifying the preventative measures to be implemented to more formal and calculative approaches that can be audited based on a generic structure to make them reasonably universal tools for assessing a wide range of companies (Hale, 2003). The arguments put forward to support this move were a combination of a systematic and verifiable way to work with safety, a management system for safety, some leeway for each company to find their own solutions to safety and work environment, a way to place responsibility to individuals (including managers and directors), and also a potential way for fostering a dynamic approach and up-to-date solutions to safety problems (Hovden, 1990).

The British Case

In the UK, beyond the Robens report, the motivation for regulators to change approaches (initially in the process industry) also came from doubts expressed by the civil society about the efficiency of regulators with a series of accidents occurring in the 1970s and 1980s (e.g., Flixborough in 1974, *Herald of Free Enterprise* in 1987, Piper Alpha in 1988), but more generally about public services as further detailed in the following section. In this context, beyond the enhancement of safety, regulators had two major challenges: to protect their liability and to demonstrate their efficacy (Power, 2004). In rail, where regulation was essentially relying on a risk assessment approach based on calculations, a similar need to change approaches was identified in the 1990s following a series of accidents in the 1980s, also more covered by media than in the past. On the decision to put a warning system on trains,

> the "Yellow book" which was their basis to perform risk assessment in the 1990s had led to calculations "proving" that it was safe enough as such. But then this series of

accidents happened. This risk assessment approach based on calculations was not working so well in rail.

(Mixed experience 1, diverse industries, January 21, 2019)

5.3.4 Civil Society: A Growing Suspicion

An interesting angle to understand the societal push toward a new approach to safety management is provided by the "risk society" theoretical perspective developed and addressed by Giddens (1990) and Beck (1992b). In this perspective, the risk is seen as amplified by the post-modern era, where risks are spreading and amplifying due to industrialization, globalization, and urbanization, created and accelerated by human activities. As described by Lupton (1999), this is one approach where risk is any phenomenon that threatens moral principles (more than anything related to modernization) and a way to keep social order, and the governmentality approach where risk is also increased by modern societies and individuals are encouraged to engage in self-regulation and be "good citizens."

In the 1980s, as described by Beck (1992b), the consciousness of self-produced or manufactured risks was increasing, creating enhanced public anxieties fueled by media. Likewise, the public defiance toward the governmental institutions and experts' opaqueness was growing at that time (Giddens, 1990; Beck, 1992b; Hutter, 2005; Power, 2004). Indeed, the public started to realize that experts were disagreeing and that governments were failing to act, not to mention a certain suspicion toward science that also contributed to modernization, thus to the development of risks. Both the public and the media were less willing to accept advice from experts or to rely on regulatory models that they suspected were lacking knowledge about a growing number of risks. The public expected decisions and demanded the right to consider decision-makers accountable (Power, 2004, p. 14). These expectations from the public were widely developed and reinforced by the public rationalist and reassuring discourse on technology to keep social order (Wynne, 1988).

5.4 An Overall Context Fostering the Convergence Toward Safety Management Systems

This section explores the overall context that made the SMS an "obvious" response to the multiple stakes and challenges of the multiple safety stakeholders. Several phenomena are reviewed because, although not directly related to safety, they provided fertile ground for an approach like the SMS to emerge.

5.4.1 An External Injunction to Justify Efficiency and Be Transparent

A big wave of deregulation occurred in the 1980s, driven by a concern with over-regulation of business and uncontrolled costs of regulation, although there were serious doubts about the efficiency of the practices in place. As mentioned earlier, the 1970s and 1980s were indeed decades of a number of major accidents, such as Flixborough (1974), Piper Alpha (1988), and *Herald of Free Enterprise* (1987). The concern with regulation and its efficiency was, in fact, much broader than safety. It led in the 1980s to the advent of the new public management (NPM) in the UK. As characterized by Hood (1995),

> the basis of NPM lay in reversing the two cardinal doctrines of PPA (Progressive Public Administration); that is, lessening or removing differences between the public and the private sector and shifting the emphasis from process accountability towards a greater element of accountability in terms of results.
>
> **(Hood, 1995, p. 94)**

Progressively, the NPM model spread across the OECD countries even though there were some variations in the extent to and pace at which the model was implemented (Hood, 1995). However, a similar wave was observed in other European countries and the US, starting in the 1980s, emphasizing the cost issue (Hutter, 2005).

> Accounting was to be a key element in this new conception of accountability, since it reflected high trust in the market and private business methods (no longer to be equated with organized crime) and low trust in public servants and professionals . . ., whose activities therefore needed to be more closely costed and evaluated by accounting techniques.
>
> **(Hood, 1995)**

Regulators were forced to legitimate their own activities by demonstrating that they were operating both efficiently and effectively—that is, without wasting resources and by proving that their activities were making a difference. They adopted a private sector style of management and risk-based approaches allowing for benchmarking public sector activities against private sector activities. These approaches not only incorporated a cost-benefit approach, but they also had the apparent benefit of being objective and transparent (Hutter, 2005). The adoption by regulators of industry risk-based tools applies to internal authorities' resource management as a perceived efficient tool to support resource allocation, tested and trusted by the business community. This growing public injunction for transparency, control, and accountability of public services was identified as a driving force for the explosion of audits in the UK in the 1980s (Power, 2000).

On a larger scale, as a generalization of the public-private partnership and the principles of the NPM, Hibou (2012) analyzes the evolution of the whole society toward increased bureaucratization, blurring the boundaries between public and private, as a way to govern by relying on norms that may not have authoritative power but that people/industries chose to comply with like ISO norms for example. As such, it could be considered inspired by the governmentality line of thought (Lupton, 1999) beyond the individual scale where industries strive to be "good industries" as citizens were encouraged to adopt certain practices voluntarily as "good citizens." As such, they engage in self-regulation like some of them did with the ISRS, before there were regulatory requirements of that sort. This evolution toward an objective control and evaluation leads to the golden age of objective, quantitative indicators and bureaucratization of activities (Hibou, 2012). One of the effects of these quantitative indicators or rating systems is the introduction of new filters through which actual practices are addressed. These categories or indicators and figures ultimately shape the way reality is thought and imagined.[11]

5.4.2 The Quality Management Era

If the origin of total quality management (TQM) can be traced to 1949, when the Union of Japanese Scientists and Engineers formed a committee of scholars, engineers, and government officials devoted to improving Japanese productivity and enhancing their post-war quality of life (Powell, 1995, p. 16), it was introduced in the US in the 1970s (when Japanese products penetrate the US market and as well as a result of the impact of Deming, Juran and other authors' writing) and in the UK in the early 1980s with the objective of enhancing product quality and, ultimately, productivity (Martínez-Lorente et al., 1998).[12] This quality management wave translated into the publication of the ISO 9000 norm in 1987 and the voluntary compliance with this standard having no authoritative power reflect the evolution of the whole society, as mentioned earlier. Quality management systems started to be part of the organizational landscape. Although ISO standards initially focused on quality, groups were set up in the 1980s and 1990s to extend the quality management approach to the environment and then safety.

The development of standards (under the lobbying of certification companies) led to a shift of focus and understanding by some high-risk companies that saw several advantages in the standards:[13] (1) they could be seen not as the minimum but as what needs to be done (i.e., something like the maximum), (2) they provided an external recognition or sign of safety consciousness, and (3) they were a cheaper option than other in-depth analysis approaches (the time needed in the field to do observations and interviews was extremely limited). Therefore, a number of industries that had adopted the ISRS initially developed at a time when there were no ISO norms took it for a management system, although it was meant to be an audit system, and welcomed more standardized approaches. "The use of the word 'system' in the ISRS questions may be the reason why there was a shortcut mistaking the ISRS for a management system" (consultant/industry 1, diverse industries, December 6, 2018). This standardization was also a good opportunity for certification companies and emerging consulting companies to make compliance with standards a juicy business and cut costs paving the way to a "low-cost safety management" (consultant/industry 1, diverse industries, December 6, 2018; Almklov, 2018).

5.5 THE DISSEMINATION OF THE SMS

Analyzing the dissemination of the SMS can be done from different perspectives. This section starts with a bibliometric look at the evolution of SMS-related publications from the 1970s to the 2000s. It then presents a more qualitative analysis of social relationships to understand how ideas traveled. This latter part is mainly based on the interview of safety old-timers as detailed in the methodology chapter (sections 2.4 and 2.5).

5.5.1 Bibliometric Perspective

A bibliometric analysis of the evolution of publications on SMS shows a dramatic increase in number from the 1990s onward, as illustrated in Figure 5.1.

Although the 1970s and 1980s decades had too few published documents to make a sound analysis of their origins (three documents in the 1970s with undefined origins; three documents in the 1980s, out of which four are of undefined origin), the analysis of the country of origin of the documents in the 1990s brings some insights as to where the idea of SMS was strongly studied and disseminated.

In the 1990s, 228 documents were listed, including 111 conference papers and 104 articles; 72 documents have an undefined country of origin. Out of the remaining ones, only three countries account for more than 5% of the overall documents listed on Scopus with the "safety management system" keyword, as illustrated in Figure 5.2.

Although slightly more diversified, especially with the emergence of China, the situation in the 2000s, illustrated in Figure 5.3, still shows the predominance of the UK and the US in the scientific literature on SMS.

Eventually, from bibliographical evidence, the UK and the US seem to have played a dominant role in the dissemination of ideas related to the SMS.

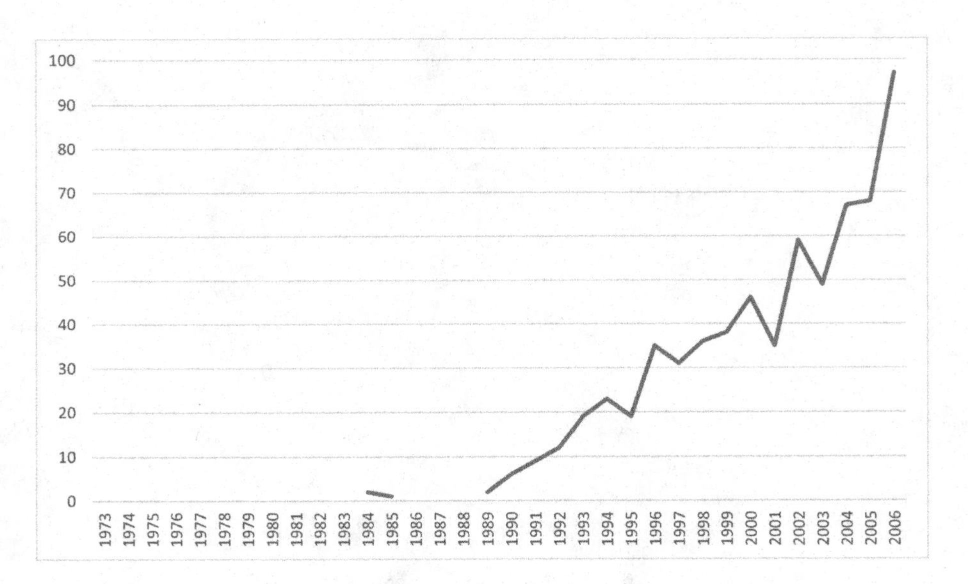

FIGURE 5.1 Number of papers for the search of the Safety Management System on Scopus.

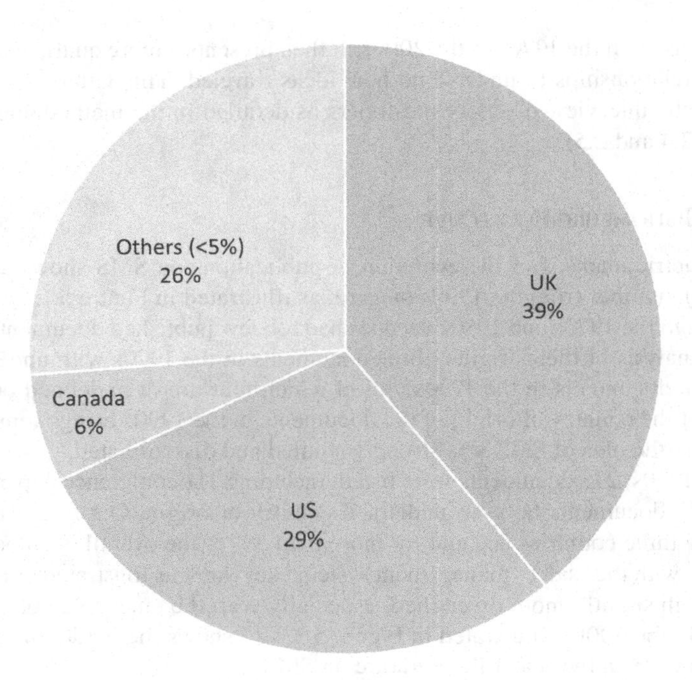

FIGURE 5.2 Origins of the documents (among known origins) in the 1990s.

Source: Scopus.[14]

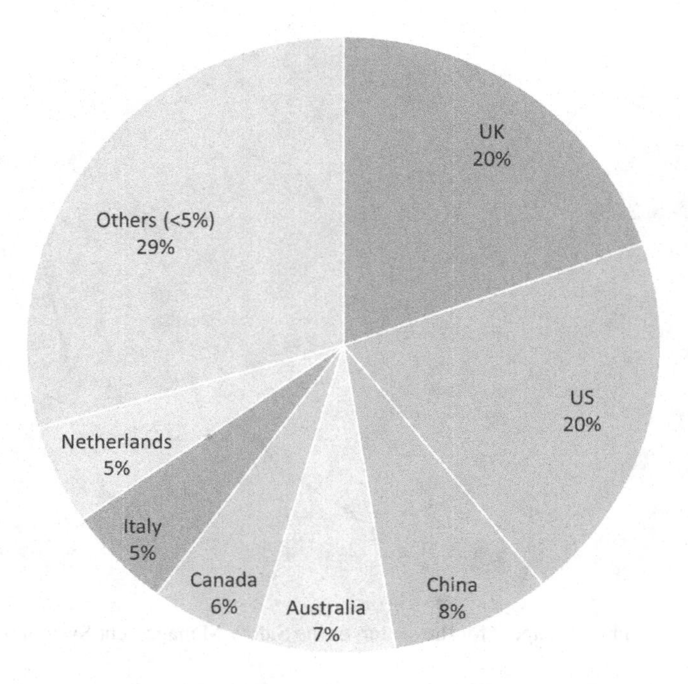

FIGURE 5.3 Origins of the documents (among known origins) in the 2000s.

5.5.2 The Circulation and Exchange of Safety Management-Related Ideas Across Communities, Industries, and Countries: How Ideas Traveled

Just having the motivation to change approaches to safety enhancement in a context where management systems became trendy and new checking mechanisms were called for by the society was not enough to spontaneously make the whole "safety world" converge toward SMSs. First of all, not every industry, safety stakeholder, or country turned to SMS. Interestingly, for those who did, it happened at different paces and places. However, in all cases, it involved a variety of social relationships allowing for ideas to travel and be exchanged from one community (especially industrials, regulators, and academics) to another, one country to another, and one industry to another. This section provides a description (not exhaustive) of the safety management discussion scenes from the 1970s to 2000.

Indeed, the social exchanges that contributed to the convergence toward the SMS started sometimes long before the SMS was officially adopted in regulatory requirements or industrial practices. Therefore, it is worth getting back in time to safety management discussion forums and channels and especially describing the dissemination of internal control, often referred to as an SMS predecessor (Hovden, 1998), or of the ISRS, another SMS predecessor (Top, 1991). From the interview of some of the key people who took part in this history of the advent of SMS, it turns out that a large variety of social relationships and networks were involved, some of them organized on purpose, others more opportunistic, not to mention personal initiatives or affinities.

The Power of Communities: A Token of Trust

Industry-Specific Communities of Practice

The need for sharing practices and learning from one another on safety led to the development of industry-specific communities of practice (Wenger, 2011) in several industries in the 1970s, as, for example, the Loss Prevention movement focused on chemical process or the 3SF learned society in France in the military domain, further described hereafter.

The Loss Prevention movement held its first workshop (by invitation) in 1972. At that time, the chemical industry was aware of the high hazards it involved. There were a series of fires as well, including the Summerland fire in the UK (1973), in which children were killed. This movement then opened up to other participants through the organization of the first Loss Prevention conference in The Hague in 1974, just before the Flixborough accident. The Loss Prevention movement was an engineers' movement, promoting a "risk approach emerging in the 1970s in the chemical industry, coming from the nuclear industry" (academic 2, diverse industries, June 14, 2019). This movement still exists, although other communities were created later in the same industry, like the European Safety Council under Dupont's initiative. This council set up in the early 1990s was a domain-specific closed forum where people could share informally, including confidential things under conditions described in a chart. Safety management was one of the main topics of discussion.

This group of 14 people met twice a year for a day and a half each time. "Every meeting involved a site visit, formal presentations and informal discussions . . . The informal exchanges were extremely rich" (industrial 2, chemistry, January 25, 2019). Exchanges also took place at a broader industry level.

> A very important event was when chemical industrials in Europe and the US agreed on a concept of codes of management practice (in the US it was the chemical manufacturers association with 180 biggest chemical industrials and in Europe the CEFIC—European Chemical Industry Council based in Brussels). They all came together and developed these codes of management practices which set forth functional responsibilities for management. . . . This was a private sector voluntary movement done because accidents (e.g. Bhopal) happened.
>
> **(Academic 4, diverse industries, June 29, 2020)**[15]

In the military domain, the 3SF (Society for the Progress of Systems Safety) was created in France in the 1970s by three industrial groups (Aérospatiale, SNPE, and SEP) to develop systems safety methods. There were monthly meetings to make progress on methods and a conference every other year (alternating with the Lambda-Mu conference) about the progress made in terms of both results and implementation.[16] "The 3SF was a 'learned society' and a conference of engineers. There were very few academics by design because we addressed industrial companies' concerns" (mixed experience 3, diverse industries, July 23, 2019).

Although it was working in the military domain, exchanges were open, including with US industrial representatives. "At that time, there was no information blockade. The experts didn't have problems sharing information except maybe numerical values" (mixed experience 3, diverse industries, July 23, 2019). The 3SF opened up to other industrial fields and to Europe in the 1990s and became the European Institute of Cyndinics. "What played a key role in this openness to other domains interested in risks were the 3SF conferences" (mixed experience 3, diverse industries, July 23, 2019).

In aviation, the first discussions about SMSs took place in the 1990s, not necessarily through identified aviation sub-communities but exclusively among aviation professionals. According to several interviewees, both from within and outside aviation, aviation was perceived as an exclusive community with very little exchange with others by both aviation representatives but also outsiders.

Eventually, forums of discussion about safety management were created through industry-specific communities interested in safety where rather open discussions could take place with a common belief that there is no competition on safety. However, another interesting characteristic of these communities is the common background of their members. Not only do they share the industrial field and the interest in safety, but they also belong to the same other community, that of engineers. These three different characteristics, though, seemed from the interviews to all contribute to creating conditions of mutual trust to exchange ideas and give credit to others. Except for the conferences organized by some of these communities that will be discussed in a later section, these forums of discussion on safety management were rather exclusive.

Communities of Users

As mentioned earlier, the ISRS, considered a major source of inspiration for SMS, led to the development of a wide community of users not only sharing the same beliefs on how safety should be managed and the same safety management tool and practices but also contributing to the development and evolution of this tool. As such, this community was a forum of discussion and exchange on safety management that is worth describing.

In 1974, Frank Bird decided to become a consultant and created the International Loss Control Institute in Loganville, close to Atlanta (US). He involved big industrials from the region in developing the first edition of the ISRS (International Safety Rating System) in 1978, thereby starting off the ISRS community that developed over the years. As such, the ISRS was a reference of best practices shared by industrial representatives and collected through yearly seminars gathering all the users. In this respect, the ISRS users were not only users of the approach but also co-developers. Several aspects played a key role in the dissemination of the ISRS and thus in the growth of the ISRS community: "professional associations and mouth to ear" (consultant/industry 2, diverse industries, December 21, 2018), as well as the commercial policy of Frank Bird. The latter led him to recruit people to develop new markets in other countries. In the 1980s/1990s, he had made arrangements with around 60 consulting companies around the world to disseminate the ISRS. In the early 1990s, the main users were located in Ontario (Canada), very close to Loganville, and in South Africa (consultant/industry 1, diverse industries, December 6, 2018). Eventually, "by the mid-1990s when the 6th edition of the ISRS came out, 6,000 companies across the world became users of the ISRS" (consultant/industry 1, diverse industries, December 6, 2018).

Members of this community were proud to display the ISRS certificate they were given by the Loss Control Institute at the entrance of the site and had a sense that the referential belonged to them, for they were involved in its development and improvement. They did not feel the need to have other interactions (with other industries or other organizations) beyond the ISRS users' seminars, making this community relatively exclusive.

The dissemination of Frank Bird's ideas and approach went through other channels, such as the National Safety Council or his lectures as a professor in the framework of the safety chair at the Georgia State University.

The commercialization of TRIPOD by Shell, as well as the setting up of the Stichting Tripod Foundation in 1998, was also a way to disseminate ideas on safety management across industries and to lead "a growing community of Tripod Practitioners and Trainers."[17,18] Yet the development of a community of users started after the emergence of the SMS.

Scientific Communities: A Common Topic, Parallel Roads

Although safety has been a concern for a very long time (records range from millennials to decades depending on the authors),[19] and despite the debates about whether it is a science or not, safety has developed as a scientific topic since the 1970s with the setting up of safety science research groups and training programs at universities (e.g., Birmingham in 1976, Delft University of Technology in 1978) or the creation of the *Safety Science* scientific journal in 1976 (known at that time as the *Journal of Occupational Accidents*). The ambition of the *Journal of Occupational Accidents*, as stated by Herbert S. Eisner in the first editorial of the first issue, goes along the lines of the development of a safety science community:

> It is in the nature of things that safety research, dealing as it does with low probabilities, is often long-winded and hence demanding in resources both human and material. It is therefore all the more important that the results of studies, wheresoever made, are available in a reasonably accessible format for the benefit of all.
>
> **(Eisner, 1976, p. 1)**

Building upon Kuhn's characterization of scientific communities and the central role played by paradigms and worldviews, one could most reasonably refer to several scientific communities interested in safety rather than a single one. This section describes some of the key elements that contributed to the development of these communities and thereby to the discussion, development, and dissemination of ideas and knowledge about safety management across different industrial fields and different stakeholders. It also discusses the persistent gap between the two main communities described hereafter—namely, (1) the reliability and safety engineering community and (2) the human and organizational (HOF) and social science community, even though they both aim to address the same problem.

The Reliability and Safety Engineering Community: Engineers Focusing on Technology Improvement

Paralleling the effort made in the military industry through the 3SF in France and its counterpart in the US, other industries with high-risk technologies further developed or refined dependability methods and techniques. These methods, mainly based on the anticipation of failures and control of risk through the (perceived) elimination of uncertainty, were disseminated and shared through two main channels: conferences and training courses.

As an example, the first Lambda-Mu conference on risk control was held in the UNESCO premises in Paris in 1978 and was organized by the CNES (Centre National d'Etudes Spatiales, space industry), the CEA (Commissariat à l'Energie Atomique, nuclear industry), and the CNET (Centre National d'Etudes des Télécommunications, telecommunication industry). As mentioned earlier, this conference involved industrials, academics, and students and alternated with the 3SF conferences, leading to a yearly occasion for people with essentially an engineering or hard science background to gather around reliability, safety, or risk management topics. Other conferences gathering similar profiles with a great overlap of participants emerged in the 1990s, such as the ESREL (European Safety and Reliability Conference) in Europe in 1992, organized by the ESRA (European Safety and Reliability Association), or the PSAM (Probabilistic Safety Assessment and Management) conferences in the US in 1991. To provide an order of magnitude, PSAM conferences gather more than 500 researchers and practitioners worldwide. The first PSAM conference, for example, included approximately 250 paper presentations.[20] Interestingly, after a few years, Lambda-Mu and ESREL organized joint conferences on a regular basis, as well as ESREL and PSAM, thereby reinforcing the international exchanges on similar approaches, within Europe and between Europe and the US.[21,22]

This community of reliability and safety engineers also developed through education or training programs on systems safety. Such programs existed already in the 1970s—for example, within ENSTA (Ecole Nationale des Techniques Avancées, a French engineering school for advanced technologies)—targeting industrial representatives wanting to develop their knowledge in systems safety and associated methods and tools. The S02 module on systems safety was created in 1976 at ENSTA. Longer education programs were also developed for students or professionals, like the specialized master's in safety and prevention of major technological risks at ECP (Ecole Centrale de Paris, another French engineering school) created in 1988. The identification of experts to give lectures or teach was performed through personal networks of these engineering schools, related either to their own former associations/learned societies like the 3SF.

Through these education and training programs, a community was developing involving people coming from different worlds and activities. "The methods [reliability, availability, maintainability, and safety methods], I developed them, and [the interviewee] passed them on to other worlds, even beyond industry, like healthcare" (mixed experience 3, diverse industries, July 23, 2019). Interestingly, in the early 2000s, in preparation for a regulatory evolution requiring the introduction of risk management in the medical world, the specialized master's in former "safety and prevention of major technological risks" at ECP became a specialized master's in "risk management in healthcare" with a common part on the fundamental methodological content. Former students of this master's program then created an association, the AFGRIS (Association Française des Gestionnaires de Risques Sanitaires), thereby developing a new community at the crossroads between scientific safety and risk management community and the healthcare safety community.

The commonality of background of the reliability and safety engineering community members also translated into a commonality of language not so much in terms of idioms but in terms of mutual understanding of the problems and needs. "In its normal state, then, a scientific community is an immensely efficient instrument for solving the problems or puzzles that its paradigms define" (Kuhn, 1962, p. 166).

Eventually, a huge community of engineers and researchers in systems safety engineering grew and developed worldwide, especially through training programs and conferences. Beyond local collaborations, practitioners and researchers gathered at least once a year at a major international conference, at which links could be created and maintained. This community shared the same background, language, and way of understanding and addressing safety. Indeed, safety is envisaged as being engineered into the system in the sense of a technical system. When the first-line operator is taken into account, it is considered a human component of which failure is called an error. Safety comes down to a technological risk control problem based on the assumption that everything can be anticipated and addressed a priori by eliminating uncertainty. The main focus of this safety science community, consisting mainly of engineers and combining industrials, regulators, and researchers, was to refine and share across high-risk industries engineering methods developed in the 1950s and 1960s, such as the failure modes and effects analysis, the fault tree analysis, and the functional analysis system technique. By its size and penetration in the industry and regulatory bodies, this community was a very powerful one, providing responses suited to the safety challenges as they were commonly stated.

The HOF and Social Sciences Community

In contrast with the HOF and social sciences scientific community,

> the issue with human and organizational factors is that there are many ideas and concepts but very few operational tools or approaches. This is why they don't get through in the industry. As an example, there is no method to assess the risk of an organizational change or it takes ten days. We don't have that time. The issue is that there is no scoring.
>
> **(Industrial 2, chemistry, January 25, 2019)**

In parallel with the development of the engineering and hard sciences safety science community, another one started developing in the 1970s involving social scientists and high-risk activities practitioners. They were interested in actual practices contributing to managing real situations beyond their anticipated modeling through procedures. The development of this community was also facilitated by education programs and conferences, as illustrated hereafter.

In Europe, Scandinavian countries were pioneers in the implementation of internal control for safety.[23] In the late 1970s, researchers had set up a Nordic Research Network on Safety that organized workshops and conferences on safety and health where they invited researchers from other countries.[24] These conferences and workshops went on and were always "a big input" (academic 1, diverse industries, September 11–20, 2018). Interestingly, there was no big plan behind the launch of these conferences, as explained by Professor Jorma Saari from the Finnish Institute of Occupational Health, who was at their origin.

> "The birth of NoFS was a pure accident.
>
> "In the late seventies, we thought there would be a need for an international safety research course, as well. I organized the first course in 1978 with good success. There were many young safety researchers on that time in the Nordic countries. . . . After the course, the Norwegian participants proposed that we should have a Nordic safety researchers' meeting. So, they organized the first meeting in 1980. This is how I learned the recipe for success. It is simple; a bunch of young researchers, warm summer in August with dark nights, a lake, sauna, and some Norwegians, especially Norwegians originating from other countries."
>
> (Professor Jorma Saari)[25]

Under the initiative of researchers including Jürgenson, Larsson, and Hovden, this approach and these conferences were spread to the rest of Europe, with the involvement of researchers from other countries, like Andrew Hale, upon invitation. It led to an annual conference (Working on Safety), going around Europe, and taking place each year in a different country. It was turned into a more traditional conference open to non-Norwegian participants without needing a personal invitation.[26]

Another scientific network started developing in 1983 at Wilpert's initiative, from Technische Universität in Berlin, with the setting up of a network of researchers and friends interested in work and technology called NeTWork (New Technologies and Work).

> At the time when NeTWork started, there was the Hacker (former GDR psychologist) model (regulation of activity model), very close to the SRK model by Rasmussen. A topic was: how do we behave at work? Rasmussen called it decision-making. How is what you do inside your head controlled? Bernhard was very critical on the first ideas of Rasmussen and Hacker. People are not on their own. They are involved in a social environment, not enough taken into account in these models. It led to the first topics of the first NeTWork workshops: new technologies and work . . . Most of the participants were psychologists in the very beginning (work and organizational psychologists) and other people interested in people at work as well as sociologists. Bernhard Wilpert was a psychologist by education but always interested in sociology.
>
> **(Regulator 5, nuclear, June 18, 2020)**

Wilpert had a lot of friends from the European projects he had been involved in (especially on the future of work and the automotive industry; one on distributed decision-making). The core group of NeTWork consisted of Bernhard Wilpert's personal network, including Eric Andriessen, Andrew Hale (psychologist by education), James Reason (psychologist by education), Jacques Leplat (ergonomist), Maurice de Montmollin (sociologist of work), Werner Ackerman (sociologist at the Center for the Sociology of Organizations in Paris, where core group meetings took place), Jens Rasmussen, and Michael Baram (a professor of law from the US, former chemical engineer, and personal friend of Wilpert, married to a German lady). In its early years, NeTWork was sponsored for the core group to meet and the workshops to be organized and take place, half by a French organization (Maison des Sciences de l'Homme, Paris) and half by the Werner Reimers Foundation (Germany). The latter foundation owned a nice location in Bad Homburg where the workshops could take place. "The core group met in Paris once a year to discuss the topic and finance . . . We met the day before the meeting, went for dinner. The core group meeting was in November and the workshop in May so we really met twice a year" (regulator 5, nuclear, June 18, 2020). The specific safety angle came later in the early 1990s. Some practitioners were also onboard, such as Koos Visser from Shell, the surgeon Bas de Mol, and Barry Kirwan from the nuclear industry, but in limited numbers. NeTWork held (and still does hold) an annual seminar, changing topics every year although, since the 1990s, all related to individual, organizational, and societal risks created by technological development.[27] NeTWork seminars involved core group members, as well as some external participants upon invitation, and were limited in the number of participants to foster exchanges, extensive discussions on the topic of the year, and networking in a friendly atmosphere. "[NeTWork] was very productive, involving more than 210 scholars from 21 countries between 1983 and 2003 only."[28] The success of NeTWork was largely due to Wilpert's personality and interest in promoting international collaboration.[29] Beyond his talent and energy to bring people together and create networks, his former colleagues insist on his personal character: "Above all, he had a genuine interest in other people's opinions and motives."[30]

According to one of the academic interviewees, Working on Safety conferences and NeTWork seminars were "the powerhouse of the safety movement, at least the safety management one (there was still a big gap with the quantitative movement)" (academic 1, diverse industries, September 11 and 20, 2018), with a big overlap in the participants in these two forums.

Beyond conferences, this community also developed through training programs and courses not only through the diversity of profiles of the students or trainees but also through the profiles of the lecturers. When setting up a course at the University of Birmingham in 1976, one of the academic interviewees had to turn to people from the industry to distill their knowledge and experience in order to include lectures on management from a health and safety perspective. Indeed, "the management experts were looking too much from a general management perspective and not specifically from a Health & Safety management one. What is important in management from a Health & Safety viewpoint was not an easy question" (academic 1, diverse industries, September 11 and 20, 2018). The interviewee eventually turned to Koos Visser from Shell (head of safety of one of the divisions), who was also involved in NeTWork and commissioned the development of TRIPOD.

With time, a new safety community developed through both conferences, workshops, seminars, and training courses focused on safety, mainly around personal initiatives to develop education programs and networks. Yet the numbers did not compare with those of the safety and reliability engineering community. Neither did the proposed ideas with the dominant way of envisaging safety. This safety science community mixed academics from various disciplines interested in humans at work, organizations, and other fields (psychology, ergonomics, social psychology, sociology, law) and a limited number of practitioners, as well as different industries and different countries. The main industries represented in this community were chemistry, oil and gas, railways, and nuclear. Interestingly, aviation had very limited participation in these safety events that were not dedicated to aviation. As for the countries that were heavily involved in this safety community, they mainly included Germany, the UK, Norway, Denmark, France, and the Netherlands. Still, this safety community was more a community of individuals than representatives of institutions, especially for the industrial members. It mainly consisted of human and social scientists, and a limited number of practitioners interested in humans and organizations developed around the core idea that what was central to any safety progress was an in-depth understanding of work situations and their dynamics. The share of voice of this community in the industrial and regulatory managerial arena was nevertheless much more limited than that of the safety and reliability engineering community, for their views were dissonant compared to existing models and practices.

A Persistent Yawning Gap

These foundational and disciplinary differences gave rise to the development of two parallel safety science communities with very few, if any, bridges between them. "Although some representatives of social sciences (around half a dozen) infiltrated 'engineering' conferences such as ESREL to try and include safety management aspects, there was never a meeting of minds between the two worlds" (academic 1, diverse industries, September 11 and 20, 2018). This gap became even more obvious with organizational aspects being highlighted and acknowledged by the second community as playing a key role in safety, whereas the first community would not integrate them as long as they could not be modeled and quantified. Beyond these conceptual differences, other aspects may have contributed to keeping these two communities apart, if not making them progressively even more distant from one another. As stated by one of the academic interviewees,

> Quantitative accountants approach satisfies managers and regulators too much. It becomes a comfortable substitute. QRAs are done by consultants cutting costs, thus they become even more superficial. More in-depth analysis costs much more and they don't see the point of it.
>
> **(Academic 1, diverse industries, September 11 and 20, 2018)**[31]

QRAs are indeed considered an acceptable means to comply with regulatory requirements even if they remain quite generic, which consultants tend to favor since it allows them to mutualize their effort. Furthermore, they do not require a deep understanding of the actual practices.

Overall, even though these communities have their own dynamics and tend to facilitate and promote the exchange of ideas and discussions within the community through conferences or training courses, other engines of transversality allow ideas to travel in other ways across different people and groups.

Engines of Transversality across Communities

Beyond the diverse communities described previously, with limited interactions with one another, other elements facilitated the exchanges between individuals, organizations, institutions, or even countries. A colloquial way of naming the three categories along which these various elements can be clustered could be affinities, money, and trauma.

Individual Trajectories, Affinities, and Personalities

Several drivers of ideas exchange and dissemination result from the personal initiatives of some individuals. They include career paths, mixed profiles and backgrounds, personal affinities triggering unexpected cooperation, or even "sell well" personalities, providing them with a significant aura, as illustrated in this section.

The career path from generally top managers was identified as a vector or dissemination of ideas from one industry to another by several interviewees. In the case of the spreading of the ISRS, for example, top managers, when leaving a company to join another one in a different domain, left with "their preferred tools, their successes. ISRS was part of them. They knew it would improve their management quality" (consultant/industry 2, diverse industries, December 21, 2018). Claude Frantzen was another example of *transfuge* from aviation to the nuclear industry, transferring ideas, especially on human and organizational factors, when he changed domains in the mid-1990s. His successor as a general inspector for nuclear safety and radioprotection (IGNSR) within EDF (French electric power utility), Pierre Wiroth, also came from aviation. They both were introduced and "'educated' to human factors and safety by René Amalberti," as one of the interviewees explained (industrial 1, nuclear, December 28, 2018).[32]

Some renowned academics were regarded as gurus with significant influencing power. James Reason, for example, with his Swiss cheese model emphasizing the role of organizations in accidents, was invited by a number of industries or conferences' organizing committees, including in the reliability and safety engineering domain, to be a keynote speaker.[33,34] One of the interviewees with a mixed profile in diverse industries, for example, after hearing his speech on his model of organizational accident in 1989 at a WEAAP conference (today renamed EAAP [European Association of Aviation Psychology]), "joined the fan club" and pushed to introduce his ideas into the ICAO Human Factors and Flight Safety working group (mixed experience 2, diverse industries, January 22, 2019). Their personality seems to play a key role in the reach of researchers' idea beyond the academic community. Whereas the book published by Barry Turner in 1978, *Man-made Disaster*, "was a revelation to a lot of people because they never thought about social factors" (academic 1, diverse industries, September 11 and 20, 2018), the reach of Turner's ideas did not compare to that of Reason beyond the academic world. However, according to several interviewees, compared to James Reason, who was a "sell well" person, Barry Turner was a shy person.

Academics doing consultancy or research interventions for industrials or regulators, in a landscape at a time (1970s and 1980s) where there was not (yet) a myriad of safety management consultants, were also powerful drivers for ideas exchange and dissemination. For example, Patrick Hudson and Wilhelm Wagenaar were appointed by Shell in the late 1980s for the development of TRIPOD.

Personal networks and affinities did play their part in bringing people together and making ideas travel. In the case of the TRIPOD development, the initiative came from Koos Visser, head of safety from one division of Shell but also part of the NeTWork scientific network. James Reason was soon brought to the project team, for Willem-Albert Wagenaar and James Reason "got along well" (academic 1, diverse industries, September 11 and 20, 2018). Another striking example is the early spreading of the ISRS in South Africa. Whereas Canada was not surprising as an early candidate for disseminating and selling the ISRS due to the geographical proximity and the word-of-mouth channel mentioned earlier, South Africa was more unpredictable. However, according to one interviewee, close to Frank Bird, it turns out that

> Frank Bird was a great hunter and went frequently hunting in South Africa (even bringing back animals from Africa for his personal zoo in his private property). By personal affinity he decided to do something in Africa. The mining industry was the natural

> domain of convergence (having also the five stars program in place that F. Bird was taking as a starting point for his ISRS performance levels definition).
>
> **(Consultant/industry 1, diverse industries, December 6, 2018)**

Funding: An Incentive to Gather Different Worlds

Another driver of transverse collaboration turned out to be the funding of research. Bringing together academics, industrials, and small and medium enterprises (e.g., consulting companies) was a condition for obtaining funding from the European Commission Framework Programs for Research and Technological Development. The funding of events, although more limited in amount, also enabled researchers to travel and exchange with other researchers from other countries.

As an example, the Framework Programs for Research and Technological Development, launched in 1984 by the European Union/Commission, helped bring academics to the industry as well as different European countries together. The first Framework Program (1984–1987) included as its sixth scientific and technical objective, "Improving living and working conditions: improving safety and protecting health and protecting the environment" (EC, 1983).

The collaboration between Wagenaar, Hudson, and then Reason with Shell also resulted from the funding by Shell of a research project to develop the TRIPOD method.

In the dissemination of ideas, this financial aspect played an important role. Talking more specifically about SMS, one of the academic interviewees states,

> In the 1980s and 1990s, that process moved across the industries. SMS was understood as the common denominator whatever the industry. The industrials would listen to academic SMS people who moved where the money was (e.g., Jim Reason moved to Health).
>
> **(Academic 1, diverse industries, September 11 and 20, 2018)**

Beyond the funding of research projects as such, funding of other types of events/work fostered the exchanges of ideas around safety management. It was the case of the conferences and meetings financed by the World Bank on "safety control and risk management" in 1988 and 1989 that brought together top-level researchers from different countries, such as Turner and Pidgeon, Rasmussen, and Westrum. The creation and perpetuation of the NeTWork scientific group mentioned earlier, gathering researchers and industrials from different horizons, was also made possible by financial support, initially from the Werner Reimers Foundation (Bad Homburg, Germany) and Maison des Sciences de l'Homme (Paris, France), today from the Fondation pour une Culture de Sécurité Industrielle (Toulouse, France, since 2007), of the yearly workshops that have made of NeTWork a great contributor to the development of safety science.

In the end, whether it was a condition to gain research credits or simply facilitate travel and exchanges through the organization of events, the funding of research was a catalyst for bringing people together around safety-management-related topics.

Accidents as Catalysts of Exchanges

In addition to the sustained channels and mechanisms through which ideas traveled and were discussed, more contextual transverse collaboration was triggered following major accidents (especially those that were covered in the media), as illustrated hereafter.

The conferences and meetings organized and funded by the World Bank were among them, as a result of a number of accidents around the world, the most remarkable one being Bhopal in 1984.

Years before, following the TMI accident, academics such as Dave Woods (cognitive psychologist) from the US went to Risø National Laboratory to work with Danish researchers.[35] "Danish nuclear was known as good and doing very interesting work. It was known as the place to go to learn nuclear safety" (academic 2, diverse industries, June 14, 2019). This collaboration contributed to continuing the work started by Jens Rasmussen on the role of humans in safety. In the aftermath of this major nuclear accident, the Presidential Commission in the US also requested the Social Science Research Council to commission social scientists to analyze the human and social dimensions of the event. The managerial challenges of operating complex, high-risk technologies were underlined in some of the papers produced in this framework (Sills et al., 1982).

Likewise, following the Piper Alpha accident in 1988, academics from Aberdeen, Scotland (close to the Northern Sea, where Piper Alpha was), such as Rhona Flin (psychologist) and Katherine Mearns (psychologist), got involved in the understanding of the accident. Through this involvement, the focus on management appeared in the Lord Cullen report (public inquiry on the Piper Alpha disaster) in 1990.

5.6 HOW DID THE SMS LAND IN AVIATION?

Although the SMS emerged and spread in many high-risk organizations, especially chemistry or oil and gas, in the 1980s and became a regulatory requirement in these industries in the 1990s, the idea of the SMS in aviation only appeared at the turn of the 21st century. This section examines the reasons why aviation initially stayed apart from the early discussions of the SMS and eventually adopted it and made it a transverse regulatory requirement across aviation activities through the issuance of the ICAO Annex 19 in 2013. It mainly builds on the interview of old-timers in safety management and, more specifically, of those coming from aviation, as well as on some historical material on aviation regulation.

5.6.1 Introduction of the SMS into the Aviation World: Preexisting Safety Landscape and Main Dates

> SMS was late in aviation. When I was first appointed to the Safety Committee of Schipol in the '90s after the El Al Boeing crashed in Amsterdam (after a decision to have an independent committee), when we asked "can you show us your SMS," it was the first time they had heard about SMS.[36] Safety was so integrated, they didn't have or see it as a separate entity.
>
> **(Academic 1, diverse industries, September 11 and 20, 2018)**

According to all the interviewees who were part of or have interacted with the aviation world, the idea of the SMS did not emerge in aviation until the early 21st century.

Historically, safety in aviation was inspired by the management of technological risks in the defense industry. "The intrinsic safety is ensured beyond intrinsically safe technology by extreme proceduralization and stringent selection and training (e.g., Apollo missions)" (mixed experience 2, diverse industries, January 22, 2019). As for the safety of technology (e.g., aircraft, navigation aids, radars), it relies on dependability studies. From a regulatory viewpoint, the leading approach was that of certification of technology and licensing of personnel. As stated by ICAO in the description of the various annexes before the issuance of the one dedicated to safety management (including SMS), "As long as air travel cannot do without pilots and other air and ground personnel, their competence, skills and training will remain the essential guarantee for efficient and safe operations."[37]

Some organizations did develop and implement safety approaches beyond regulatory requirements. For example, line-oriented safety audits (LOSAs) were developed following an initiative pushed by Delta Air Lines (an American airline) in the mid-1990s to gather data from the cockpit in normal operations on safety strengths and weaknesses. These data were meant to be used as inputs for airline-specific safety analysis. Other airlines followed on a voluntary basis, and ICAO eventually supported LOSA and published a LOSA manual (ICAO, 2002), which further promoted the approach. Yet nothing like the SMS was properly pushed by aviation operators but rather came from regulatory bodies.

ICAO issued the first Safety Management Manual as such in 2006, and Annex 19 was fully dedicated to safety management (safety program at the state level and SMS at the aviation organization level) in 2013. Yet previous annexes already included safety management SARPs (standards and recommended practices) for service providers. These standards became progressively applicable to different aviation organizations, from 2006 for Air Traffic Services and certified aerodromes to 2013 for aircraft design and manufacturing organizations.[38]

In the US, the FAA launched an SMS initiative in 2003–2004 and published its first SMS guidance for air operators in 2006 (AC-120-92).

Interestingly, in Europe, Eurocontrol had issued the ESARR 3 (Eurocontrol Safety Regulatory Requirement) applicable to its member states on the use of SMS by ATM service providers in 2000.

> Initially it was a sunshine regulation, i.e., it was not mandatory but was made known to peers (which was a strong incentive to have it). It became mandatory in the framework of the Single European Sky 1 (2000) that was ratified in 2003/4.
>
> **(Regulator 3, aviation, February 20, 2019)**

5.6.2 How Come the Notion of SMS Did Not Reach Aviation Earlier: A Closed World

Several sources of evidence converge to describe aviation safety as a closed world.[39] Interviewees from both aviation and outside aviation note the very limited (if any) interactions of aviation with other fields and the challenge it is to discuss aviation safety when not an aviator.

As stated by one interviewee coming from outside of the aviation world: "The mindset in aviation was like in the railways: you have to have been a first-line operator for a long time before someone would listen to you about safety" (academic 1, diverse industries, September 11 and 20, 2018). Not only was aviation a closed world, but safety was a matter of operational people. Considering the emphasis on training of operational aviation personnel and the associated licensing, bridges between other industries and aviation were not easy in terms of career paths. At the managerial level, the situation was similar since "managers come from the ranks" as underlined by an interviewee having worked in different countries and organizations in aviation (regulator 2, aviation, September 25, 2018). As for more institutional cooperation and exchanges, interviewees from aviation share the view that it was mainly limited to other aviation actors rather than other industries.

Although silos do exist within aviation as well, especially between industrials, researchers, and administrations, some initiatives are taken to reach beyond them. It is the case of the Eurocontrol-FAA joint research seminar that first took place in 1997. This was an attempt to bring together people doing research on ATM worldwide, whether from research centers from administrations, institutions, or universities. The seminars were put together by Jean-Marc Garot, head of the Eurocontrol Experimental Center, and Jack Fearnsides, director and general manager of the Center for Advanced Aviation Systems Development of the MITRE Corporation. This initiative finds its origins in the career path of Jean-Marc Garot. Indeed, as stated by one of the interviewees, Jean-Marc Garot "had been seconded to the FAA for two years (1985–1987). He became close to the people who became top managers of the various US entities in charge of research" (regulator 3, aviation, February 20, 2019). These seminars were an occasion (and still are) for creating relationships within the ATM world. "These conferences brought together all the research centers from institutions and universities doing research on ATM" (industrial 3, aviation, February 12, 2019). Back in the late 1990s, these seminars gathered around 100 attendees (now it is between 200 and 300). "It is a conference by invitation, not an open one. Yet, should anyone want to attend it, as long as (s)he had a connection with ATM, (s)he is invited. The industry is attending regularly" (regulator 3, aviation, February 20, 2019).

In addition to this formal initiative, some institutional cooperation exists, for example, between Eurocontrol and the tech center of the FAA and NASA Ames in the US, but they remain within the aviation world. Similarly, some education programs are designed for current or future aviation professionals, such as the degree in human factors at Paris 5 University in France, co-organized by the Paris 5 University and the IMASSA research center (a military research center of human factors), or the master's degree in human factors and system safety at Lund University in Sweden. However, relationships with academia at large, apart from aviation specialized departments, are very limited and are based on the personal relationships of employees.

Most exchanges remain within the aviation world, including between different aviation organizations involved in different aviation activities. "Within aviation, we have always had interactions and collaborations with manufacturers . . . We don't really succeed to develop links with other industries, even transport ones" (regulator 3, aviation, February 20, 2019). "I attended conferences of British pilots and ATCOs" (industrial 3, aviation, February 12, 2019).

The analysis of major general safety conference programs corroborates the perceptions and the observation of aviation as a world closed in on itself. For example, in the case of the PSAM (Probabilistic Safety Assessment and Management) conferences, which were major safety events gathering hundreds of industrials, regulators, consultants, and academics, the representation of aviation was almost nonexistent.

At the first PSAM conference in 1991 in California, US, several industrial fields were represented through presentations, especially oil and gas, chemical, nuclear, defense, space, and construction. Out of the nearly 250 presentations overall that were given during the conference, only one session, entitled "Aircraft Accident Analysis and Systems," could relate to aviation. However, this session included only one presentation on Bayesian methods for aircraft accident probabilities by Mosleh and Kaplan, experts in Bayesian methods but not coming from aviation. As such, it was not the world of aviation that participated in the conference but rather academics that worked on aviation data. At the second PSAM conference in 1994, two sessions were dedicated to transportation out of 108 sessions in total. However, only three presentations were related to aviation out of over 330 presentations overall. Two of them were rather technical (from Israel and the US), and one from the NATS (UK National Air Service Provider) was on system safety management.[40]

One rare but notable bridge though between aviation and the outside world is the collaboration launched in 1996 by the FAA with the Sandia National Laboratory, a US research organization specialized in safety and reliability, especially in the nuclear industry, in order to revisit their oversight approach.

> We use applied system safety principles to identify the critical elements of a system, identify the hazards and the risks associated with them, and then address those in a proactive manner. We've been able to lend to the FAA our expertise in system safety and quality improvement based on decades of applying those principles to high-consequence engineering problems.
>
> **(See www.sandia.gov/LabNews/LN05-22-98/faa_story.html)**[41]

5.6.3 Motivations to Change Approaches: A Regulators' Concern above All

In US civil aviation, according to Stolzer and Goglia (2015), the need for a change in safety approaches was anticipated in the 1980s/'90s with the flattening of the accident rate and the projection of traffic growth that would lead to an increase in the number of fatal accidents, unacceptable to the public. A chart put together by Boeing and shared among the aviation community showed that the curve of the accident rate evolution had flattened over the two decades from mid-1970s to the mid-1990s. This accident rate, combined with the anticipated traffic increase, would lead to major fatal accidents nearly every week in the following decade, as underlined in the presentation made by Boeing, a message repeated by the FAA administrator (Walters,

2000). The number of fatalities in 1994 in the US (264 versus 18 in 1993 [source: NTSB]) triggered a reaction from the secretary of the Department of Transportation in January 1995 with the organization of the first Aviation Safety Summit, gathering "950 representatives from airlines, unions, manufacturers, and regulators . . . to discuss the formulation of new initiatives to improve safety and increase public confidence" (Stolzer & Goglia, 2015, p. 55). It led to voluntary airline safety programs (e.g., Advanced Qualification Program, a training program customized to pilots' proficiency; Line Operations Safety Audits, observation of line flights to collect safety-related data on pilots' performance, environmental conditions, and operational context to be used as inputs for an airline-specific safety analysis). "Keeping these programs voluntary rather than mandatory has been a central goal of the industry throughout the decade of negotiation with the FAA" (Stolzer & Goglia, 2015, p. 59). On the regulatory side, a pivotal event was the ValuJet accident in 1996 and the shortcomings identified in the FAA oversight.

> Administrator David Hinson, testifying before Congress a few weeks after the crash, identified two mistakes that the FAA had made relative to ValuJet: failing to understand and deal with the effects of rapid growth and failing to foresee the difficulties created by virtually complete outsourcing of maintenance.
>
> **(DOT, 2008, p. 28)**

This crash pushed the FAA to develop a new oversight approach called ATOS (Air Transportation Oversight System). However, the motivation to change safety management approaches went beyond the oversight framework. The FAA realized that in order for system safety to be efficient, it must be practiced by the system owner—that is, the operator (Stolzer & Goglia, 2015).

In Canada, the pivotal event to change regulation and oversight approaches, at least for airlines, was the Air Ontario accident that occurred in 1989 and, more specifically, the public inquiry that followed. Airline deregulation had been introduced in Canada two years before, in 1987, by the National Transportation Act. The objective of the public inquiry launched by the Canadian government was "to determine how organizational factors may have contributed to the accident and, more importantly, to determine how the industry's unhealthy economics and poor safety culture were becoming a serious threat to the traveling public" (David-Cooper, 2002, p. 35). After two years of investigation, the report concluded that organizational aspects contributed to the accident, an unprecedented finding in an aviation accident investigation. The broadening of the scope of the dimensions that could affect safety to organizational characteristics called for a new regulatory and oversight approach to address them.

In Europe, regulation was historically also essentially based on certification. The need for moving toward new regulatory approaches came just before the turn of the 21st century, with the corporatization of ANSPs in some countries. For example, NATS became a wholly owned subsidiary of the UK CAA in 1996. "The privatization or corporatization of public services was a strong incentive to rule the economic and safety aspects independently from one another to . . . make sure that safety wouldn't be sacrificed to the benefit of other business indicators" (regulator 3, aviation, February 20, 2019).

Conversely to the pioneer fields in SMS, there did not seem to be a push or a particular enthusiasm from industrials to move toward a new approach based on quality concepts. Evidence for motivations to change safety approaches essentially comes from regulatory bodies in aviation triggered by either the anticipated need to enhance safety to "compensate for" increases in traffic and keep absolute numbers low enough to be acceptable to the public, overcome some shortcomings of the existing approach, or a change in the industry landscape, calling for a clarification of control functions.

5.6.4 How Did the SMS Make Its Way Through: Internal Forces, Tenuous Bridges between the Closed World of Aviation and the Outside World, or Both?

How the SMS eventually landed in aviation does not seem to bear an easy answer or at least a similar answer all around the world. To some, it is still a mystery. As stated by one of the interviewees, "What is surprising is how people managed to introduce it [the SMS] in aviation since it bothered everyone: Managers since it meant additional work and responsibility; Operational people (e.g., pilots, ATCOs, engineers) since they were convinced they were controlling their work, thus safety" (mixed experience 2, diverse industries, January 22, 2019). However, from a regulator's viewpoint, "Nothing happens in aviation unless there's regulation behind it" (regulator 2, aviation, September 25, 2018). It seems, indeed, that the SMS was introduced to aviation by regulators more than by aviation operators. Yet how the SMS made its way through to aviation regulation remains a question.

The scenario seems to differ between Europe and North America. Some European interviewees from aviation consider that the SMS was imported to aviation from other industries. In North America, the assumption of an injection of quality concepts into safety by aviation regulators seems to be preferred.

In European ATM, with the corporatization of some ANSPs, Eurocontrol created two entities in 1998 in order to ensure some independence from service provision and safety regulation: the Safety Regulation Commission (SRC), supported by the Safety Regulation Unit (SRU), which is responsible for the development of harmonized Eurocontrol safety regulatory objectives and requirements for the ATM system within the ECAC (European Civil Aviation Conference) area.

> The SRC was consisting of Commissioners who were regulators from ECAC countries, and when the state and the operators were not split, the countries were represented by a pair (one representative from the state and one advisor e.g. one person from DFS in Germany, DFS being the advisor of the German MOT).... There were approximately 70 to 80 people in the SRC, around 2 persons per country. . . . The SRC didn't have exchanges with academics. We discussed with ANSPs.
>
> **(Regulator 4, aviation, February 25, 2019)**

Since the NATS (UK ANSP) had become a fully owned subsidiary of the UK CAA in 1996, the UK already had the experience of the split between the entity in charge of ATM service provision and the entity in charge of rule-making and oversight—that is, the experience of a regulatory interface.

> How this interface would work didn't come from aviation in the UK. It came from the experience in the railway industry after it was privatized in a context of a liberal political system and more generally a certain vision of the world. The UK experienced a number of rail accidents which led to the clarification of control functions, especially for safety.
> **(Regulator 4, aviation, February 25, 2019)**

This UK experience, initially gained from the railway industry, turned out to be very influential in European ATM regulation, as confirmed by all the interviewees from European ATM.

> The UK was perceived as dominant. The chairman of the SRC came from the UK. The head of the SRU came from the UK. The other countries didn't have the experience. They were not doing such thing in their countries. . . . The UK Civil Aviation Authority Safety Regulation Group had a vision of how the regulator should intervene, considering that they had an issue to remain competent on technology and that they didn't want to take the responsibility of a detailed prescriptive and oversight approach. They were willing to focus the approval on the safety management system rather than on operational aspects. They had had bad experiences.
> **(Regulator 4, aviation, February 25, 2019)**[42]

Mid-2000, Eurocontrol issued a safety regulatory requirement: the use of SMSs by ATM service providers.[43] As reinforced by one of the European interviewees, "Eurocontrol is an international organization. Therefore, it is pulled by the countries that may be ahead of others. It was the case for the SMS" (regulator 3, aviation, February 20, 2019). However, despite the "early" developments of safety management in European ATM, the ESARR 3 was initially just a "sunshine" regulation—that is, there were incentives to implement it, but it was not mandatory.[44] The real push toward a change in regulatory approach happened following the Milan Linate and Überlingen accidents, in 2001 and 2002, respectively, revealing that there were "organizational and oversight shortcomings" (industrial 3, aviation, February 12, 2019). The SMS became mandatory in European ATM in the framework of the Single European Sky that was ratified in 2004.[45] At that time in Europe, there was no SMS requirement for other aviation organizations beyond ATM (e.g., airlines, airports, manufacturers).

Following the Moshansky Commission report on the Dryden accident investigation and the recommendations that were put forward, Transport Canada reviewed its safety regulatory approach. They

> were doing something bringing together organizational aspects, business management and quality management as a kind of new accident prevention program. They took up the basic notions of safety (hazard, risks) and thought that by adding a quality management approach, the industry could use their resources more efficiently. Their approach was very influential because they were the only ones to have a structured approach. They proposed a framework of 17 components of SMS. It was an ABC, a beginning.
> **(Regulator 2, aviation, September 25, 2018)**

The initial SMS developed by transport Canada is presented in Figure 5.4.

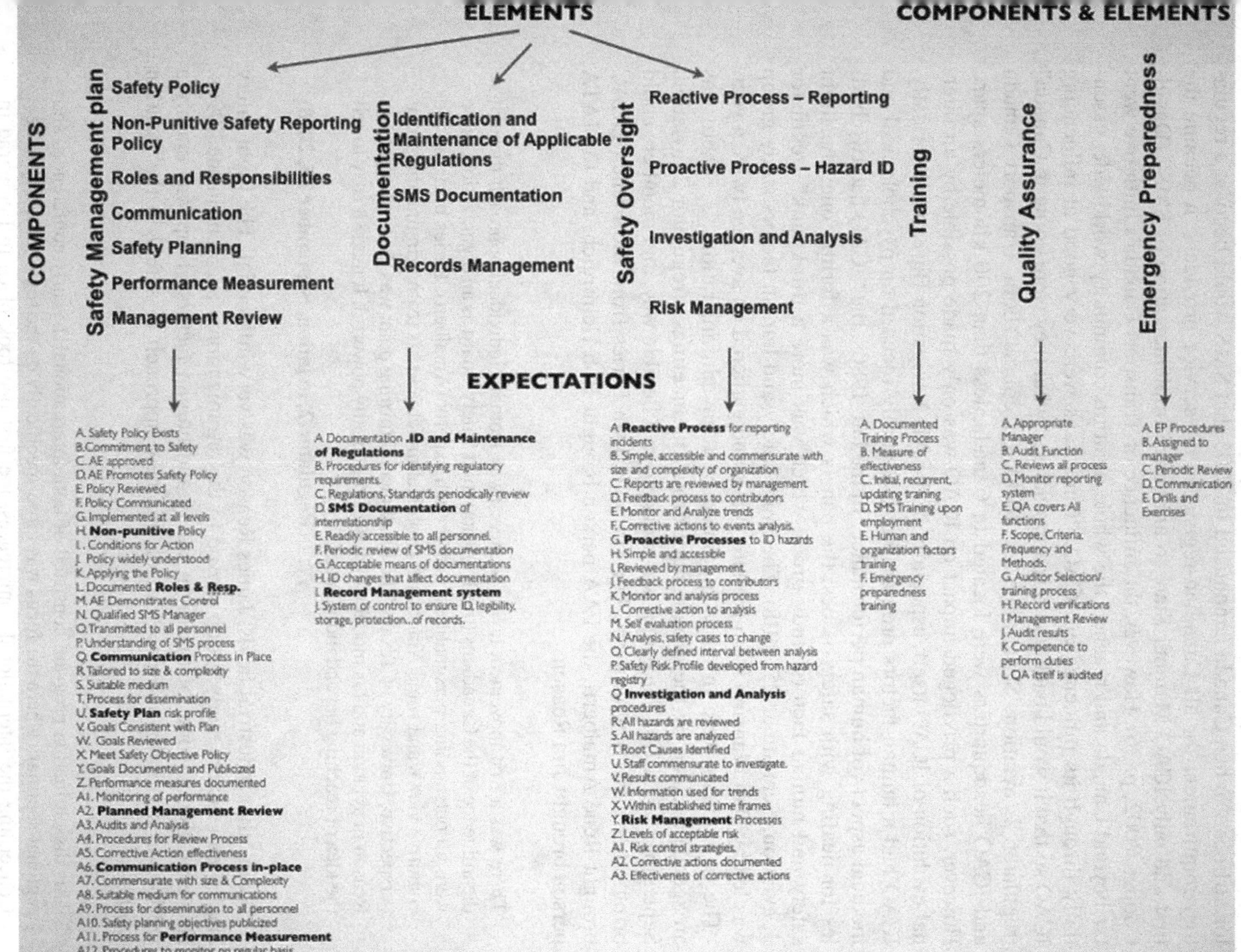

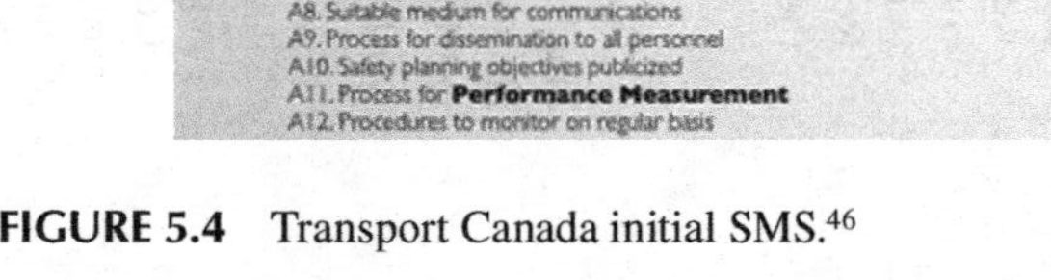
FIGURE 5.4 Transport Canada initial SMS.[46]

Eventually, Transport Canada became a pioneer in the development of the SMS for airlines. "When Canada announced in 2005 that SMS would become a regulatory requirement for all Canadian-registered air service providers, it became the first and only ICAO Member State to announce implementation of SMS" (David-Cooper, 2002, p. 36). However, the influence of Transport Canada's initiative went far beyond Canada and reached the whole aviation community worldwide, essentially through its influence on ICAO. "Canadians were very good at influencing ICAO so that it would back up their approach i.e. what they were doing in Canada" (regulator 2, aviation, September 25, 2018). Of course, both Transport Canada and ICAO headquarters were located in Canada, less than 200 kilometers apart. However, such a privileged impact on ICAO was only made possible by an internal evolution of ICAO itself, especially of the Air Navigation Bureau. Historically, ICAO had a human factors program but essentially focused on psychological and medical aspects of human performance. In the late 1980s, the "ICAO had two bodies for dealing with safety issues: the panels which was a formal one; the flight safety and human factors study group which was more a think-tank" (regulator 2, aviation, September 25, 2018). The flight safety and human factors study group was led by Dan Maurino from the late 1980s and gathered on average twice a year. "The individuals were initially appointed by their administration . . . There was an agenda decided by the group and it drove the meetings" (regulator 2, aviation, September 25, 2018). The group was working in a flexible way, and members could bring other experts with them. For example, Jean Pariès from the French DGAC brought René Amalberti, the FAA person brought Bob Helmreich, and the IATA person brought Jim Reason.

> There was a first movement trying to steer away from an individual approach through the influence of the Canadians (the Transportation Safety Board as an organization, there were no outstanding personalities) and French experts (Amalberti, Pariès bringing the cognitive view), and some Americans (Bob Helmreich and his team bringing the social perspective, University of Texas). The fundamental turning point was the landing of Jim Reason in aviation and his model which was graphically powerful. It made it easy to sell the idea of reaching beyond individuals.
>
> **(Regulator 2, aviation, September 25, 2018)**

As in other industrial fields, James Reason was very influential. The flight safety and human factors study group moved to the organizational accident in the 1990s. However, the outcome of the flight safety and human factors group was mainly a worldwide dissemination program of this new approach to safety in the aviation world.

> This dissemination program consisted of both international 2-yearly conferences and regional seminars where the focus was more specifically on the regional safety issues. Concerning the international conferences, the first one took place in Leningrad in 1990, the last one in Santiago (Chile) in 1999. These conferences involved on average 400 to 450 participants from the industry (manufacturers, CAAs from major aviation countries, airlines, MROs, ATO). The speakers were a mix of a relatively stable core group involving James Reason, Ron Westrum, Bob Helmreich, Neil Johnson, Earl

Wiener, Jean Pariès and speakers from the country/region where the conference was held.

(Mixed experience 2, diverse industries, January 22, 2019)

Although in the mid-1990s the flight safety and human factors study group started to import management aspects in safety, at the time when quality management was introduced in aviation, it seemed too early in the aviation world in terms of mindset. What the industry was looking for were rather applied tools to integrate human factors. At ICAO as well, the idea of the SMS was first brought from another industry.

> The first time someone came to ICAO with the idea of managing safety instead of preventing accidents was Brian Humphries who was the Director of aviation safety at Shell. They had imposed quality standards to their subcontractors and management. Initially, I didn't get the idea, but we maintained contact. They invited me to meetings in London. They were looking at the low-level events, the low hanging fruits and context. Things started clicking. Organizational aspects highlighted the importance of context as well. Things started coming together, but still with the aim of preventing something from happening. We hadn't made the jump to normal operations and something beyond preventing accidents. It didn't happen until the turn of the century.
>
> **(Regulator 2, aviation, September 25, 2018)**

As mentioned earlier, the main outcome of the flight safety and human factors study group was a dissemination program. Although the group operated for 15 years, it barely led to any new standards, just a few amendments to ICAO annexes. This philosophy changed with the appointment of William Voss as Head of the Air Navigation Bureau in 2004.

> He was the person who really forced (at least me) to go beyond what we were doing in HF with the different disciplines and to inject some quality into the process. He told me: get your buddies together and find a way to convince organizations to reformulate their budget on safety, i.e., to allocate their budget based on business process. Don't think of accidents as a parameter.
>
> **(Regulator 2, aviation, September 25, 2018)**

The change also affected ICAO's way of working on safety and human factors. First, a safety management program was created in replacement of the international flight safety and human factors group with a mission to support regulation—that is, the development of ICAO standards (conversely to the flight safety and human factors study group). Second, the safety management program was an ICAO internal project (no panel, no group) with hardly any exchange with other countries or entities. "There were occasional contacts with Canadians but that was all" (regulator 2, aviation, September 25, 2018). It was the time when Transport Canada was working along these lines as a follow-up to the recommendations following the Air Ontario accident in Dryden.

Eurocontrol also worked with ICAO, but rather the Air Navigation Commission division of ICAO. "Eurocontrol and ICAO had a cooperation agreement. We met on a regular basis. We participated in the ANC. I did a presentation of the ESARRs at the ANC" (regulator 4, aviation, February 25, 2019).

Eventually, around the mid-2000s, the SMS landed in ICAO. The will of ICAO was then to quickly issue standards and refine them based on the feedback of the users, namely, safety managers from the industry and regulators from national CAAs in charge of writing the national regulation. The way the safety management program operated was totally different from the flight safety and human factors study group. As evidenced by an interviewee:

> With the HF work, we had a strong basis of knowledge. It was not the case with Safety Management. It hadn't been tested. There were two options for ICAO:
>
> - Start doing research and collect information and don't do anything until we have all that we needed.
> - Start with a light frame of reference and learn by doing, then update and upgrade it.
>
> We chose the second option.
>
> ICAO started a program of SMS seminars (around three seminars per week, well over 100 seminars in total). We presented what we had in terms of guidance material and collected feedback. The questions were: Is it clear? Understandable? Does it help you? We got a lot of feedback. We developed 14 versions of the SMS seminar material by integrating this feedback. There were huge improvements. Eventually, it led to a second round of standards published in 2007.
>
> **(Regulator 2, aviation, September 25, 2018)**

An Industry-Specific Context That Makes Aviation a Late Follower of the SMS Trend

Aviation was not part of the pioneer industries in the SMS genesis. It was rather a late adopter of an already well-established approach, a follower of other high-risk industries that developed it and adopted it much earlier. Conversely to most SMS pioneer industries (especially chemistry or oil and gas), aviation safety was historically based on the intrinsic safety of technology completed by an extensive proceduralization and highly trained operational personnel. Additionally, the management functions were ensured by former operational personnel, making it an industry closed in on itself with very limited interactions with other industrial fields. The drivers for a change of safety management approach in aviation were mainly transformations of the industry itself, whether through the corporatization of ANSPs, deregulation of commercial flight services, or the anticipation of dramatic traffic growth, thus risk exposure, even though some major accidents acted as catalysts. Although presented as a promising novel safety management approach from the mid-2000s in aviation, the SMS was not a new approach but rather the heritage of an earlier industrial and societal context.

5.7 CONCLUSION: THE SMS AS AN EMANATION OF THE TIME AND CONTEXT WITHIN THE INDUSTRY AND WAY BEYOND

Although the SMS is claimed to be an approach to enhancing safety through its management, it turns out that the emergence and spreading of the SMS may be best understood as an emanation of the time and context. Investigating the SMS origins suggests that it is worth trying to understand it from a coupling perspective and apprehend it in its "full ambiguity and complexity" (Gladwell, 2019, p. 290). Through the elements presented previously, this section looks at the SMS through coupling lenses and explores to what extent it is coupled with a time and a context. More specifically, it is coupled with the stakes of the various stakeholders at the time but also with the response to these stakes that were adapted to the time.

Considering the societal context in the UK and the associated regulatory crisis leading to the NPM and its dissemination in OECD countries in the 1980s and 1990s, the SMS provides a solution to the regulators that is convenient in many respects. It reduces the resources needed to perform the oversight mission by limiting the scope of oversight to checking the organizational arrangements (most often the structural ones only) put in place by high-risk companies to manage safety. Conversely to a monitoring of operational activities, this approach requires less time, especially on-site, and expertise in operations. Interestingly, the resource aspect, both qualitative and quantitative, was also an issue in Norway, even though the initial main driver was not a societal push for more efficiency of public services but rather a competency and capability concern (Hovden & Tinnmansvik, 1990). In addition, by building on an approach already used by industrials and considered efficient by them to support resource allocation on safety (namely, internal control), the SMS reinforces the evidence of efficiency. Indeed, even though there is no scientific basis for the SMS, the civil society at that time had high trust in private business and high defiance against civil servants. According to Hood (1995), the stress on private sector style of management practice is one of the dimensions of change introduced by the NPM. In this context, the SMS became a self-evident good safety management approach (Hovden & Tinnmansvik, 1990). Besides, the risk management approach at the core of the SMS responds to the societal call for more transparency by making the risks explicit, as well as their criticality and mitigation measures where needed (Hutter, 2005). For regulators, this apparent transparency reaches beyond the sole objective of legitimating their activities. It also constitutes an alibi, a demonstration that everything was done to prevent failure. Moving away from detailed prescriptions and regulations specifying the measures supposed to prevent unwanted events and adopting a monitoring approach of industries' self-regulation induce changes in responsibilities. As such, it is a form of preventive accountability to avoid being blamed in case of failure (Power, 2007). Incidentally, the phenomenon analyzed here with a focus on safety is not specific to safety. The emergence of risk management and associated management systems occurred in areas other than safety, where internal control systems used by private sectors to increase their performance became regulatory objects (Power, 2004). More generally, the move toward the risk management of everything is a way to define a structure of apparent control and accountability that provides the reassurance called for by the public, including regarding the governance of the unknown (Power, 2004).

For industrials, beyond the stakes associated with the reduction of accidents, especially ethical and cost-related ones, the SMS helps to improve the organizations' overall performance in the context of increased competition (Martínez-Lorente et al., 1998). Besides, it meets a variety of imperatives within a number of industrial organizations shared by industrial and consultant interviewees, such as being a practical tool, being relatively quick and easy to implement (at least if limited to the customization of existing supporting material), providing both internal and external recognition, and allowing for measuring results through quantitative indicators. All this in a context where management systems built on indicators were becoming trendy, if not best practices, following the quality management move. As summarized by an interviewee, "It was a trend. Everybody wanted it" (mixed experience 4, oil and gas, September 17, 2020).

In this context of efficiency, transparency, performance, measurement, and practicality, the scientific communities developing in parallel in safety science did not play the same role or have the same influencing power. Although the need for considering organizational aspects in order to enhance safety was introduced by the social science community, the way this need translated into practice was largely shaped by the engineering community of safety science. As an extension of the standard reliability and engineering method, relying on the same conceptual model, the SMS was chosen as a "comfortable" proxy for accounting for the organizational contribution to safety. This engineering community developing and promoting reliability, safety, and risk management approach was not only the oldest one to take an interest in safety but also consisted of a large part of industrial and regulators' representatives. The HOF and social science community interested in safety was much newer and consisted mainly of academics and some individuals from industry or regulators personally interested in this perspective. In this landscape, the reliability and safety engineering community, in a kind of circular reinforcement, had both a better penetration in the industry and regulatory landscape and a more appropriate response to the industry, regulatory, and societal requirements of efficiency, transparency, practicality, objectivity, and measurability.

In this context of influence imbalance, all the literature on organizational aspects (e.g., organizational blindness, normalization of deviance, or decision-making processes adapted to the expertise required by the situation) was translated into two main elements: the introduction of organizational factors in hierarchical control models and the development of a management system for safety to support managers for allocating resources to safety. Although this decoupling between organization science and high-risk activities has been extensively analyzed by Bourrier (2017), it reflects the coupling of the SMS with the context of ideas where the interactions between the engineering and the social science communities working on safety were very limited and their respective share of voice within the industry or regulatory bodies were uneven.

Despite the new insights on safety and exchanges that were developed by the HOF and social sciences community in the 1980s and 1990s, the SMS remained mainly based on past concepts and understandings. Could it have been different? From a coupling angle, showing the diversity of aspects at stake for different stakeholders and the overall context, including the power imbalance between safety science communities, it turns out that the SMS was an emanation of the time and context.

NOTES

1. That is, at the crossroad between satisfying and sufficing.
2. Interestingly, this was not always the case. "In 1928, the American Engineering Council's Committee on Safety and Production conducted an extensive study of the

association between safety and production efficiency. The conclusion was drawn generally that the safe plant is an efficient plant" (Grimaldi, 1965, reprinted in 2006, p. 57).

3. The relationship (or absence of) between the members of these three different schools of though is addressed in a later section (0).
4. With the underlying assumption that there can be an ideal organization that if working according to its processes would be intrinsically safe.
5. Distinction used by Hale and Hovden (1998).
6. Note: Although Weick and Sutcliffe (2015) took the HRO theory one step further toward an explanation of how HROs better manage the unexpected than other organizations, these developments were not yet published at the time the SMS spread across a variety of industries (in the 1980s and 1990s), including when it reached aviation. Likewise, the whole resilience engineering school of thought appeared long after the SMS emerged and started being adopted in many high-risk industries. The first publications (around the mid-2000s) coincided with the adoption of SMS by aviation and cannot be considered part of the SMS background landscape.
7. Although commonly known for the safety pyramid inspired by Heinrich's work as well, Frank Bird Jr. also spent a significant part of his career working on management practices supporting safety. One of his major contributions in this area is the development, improvement, and dissemination of the International Safety Rating System (ISRS) to support companies in their organizational and managerial arrangements and practices contributing to safety. The ISRS is considered one of the sources of inspiration of the internal control and the SMS, as further explained later.
8. The ISRS will be further developed later.
9. Health, safety, and environment.
10. The internal control consisted of regularly checking the organizational arrangements meant to ensure the control of safety risks, mainly if not exclusively from an organization structure viewpoint. As an illustration, these arrangements would include the existence of safety committees and safety policy (still found in SMS requirements today) deemed necessary to manage safety. Despite the absence of scientific evidence of its actual positive impact on safety and the mixed feelings it induced to industrial safety stakeholders (Hovden, 1990), internal control was commonly adopted as a good practice for safety management. The idea of internal control dates from the late 1940s and originated from the financial domain to improve financial control and reporting quality to strengthen the accountability of boards to shareholders (Spira & Page, 2003).
11. This phenomenon was already referred to earlier when discussing safety indicators and the tendency of organizations to manage safety indicators rather than safety practices.
12. Juran, J. M., & Wescott, M. E. (1948). *Control Charts: An Introduction to Statistical Quality Control.*

    Deming, W. E. (1943). Opportunities in mathematical statistics, with special reference to sampling and quality control. *Science*, 97 (2514), 209–214.

    Deming, W. E. (1950). To management. In speech at Mt. Hakone Conference Center, Japan.
13. For example, Bureau Veritas, DNV (Det Norske Veritas).
14. Average population in the 1990s: UK, ~58 million inhabitants (source: Office for National Statistics, UK); US, ~264 million inhabitants (source: US Census Bureau); Canada, ~29 million inhabitants (source: StatCan).
15. Historically in the US, big industrial associations were created after the advent of federal and state safety regulation (from the 1940s to the 1960s): "Companies moved ahead and started to promote private technical standards and recommended practices as a way of foreclosing too much governmental intervention in their lives" (academic 4, diverse industries, June 29, 2020).
16. The oldest and largest international conference on reliability and maintainability (more generally on risk management, Lambda and Mu being the name of, respectively,

reliability and maintainability symbols) organized in France. In 1994, Lambda-Mu and ESREL (European Safety and Reliability) organized a first joint conference.

17. TRIPOD is an accident and incident investigation method that takes into account organizational factors based on James Reason's so-called Swiss cheese model (Reason, 1990).
18. See https://publishing.energyinst.org/tripod/home.
19. Hale and Hovden (1998), for example, refer to measures (strong incentives to manage one's work safely) two millennium BC in Babylonia whereby should a house fall down and kill the owner, the mason would be killed, and so would his son, should the son of the owner be killed.
20. See www.iapsam.org/psam-mtgs.html.
21. The first joint Lambda-Mu–ESREL conference took place in 1994 in La Baule, France.
22. The first joint ESREL-PSAM conference took place in 2004 in Berlin, Germany.
23. The first definition of internal control dates back to the middle of the 20th century and comes from the financial and accounting fields (Cordel, 2019).
24. The first NoFS (Nordisk Forskningsnetværk for Sikkerhed) conference took place in 1980.
25. Document on the Jubileum of NoFS-1980–2010: https://nofs.net/wp-content/uploads/2013/04/JUBILEUMSSKRIFT-1980-2010.pdf.
26. The first open Working on Safety conference took place in 2002 in Elsinore, Denmark. The Working on Safety network still exists and continues to organize conferences (www.wos2021.net/about-wos-net).
27. See https://network-network.org.
28. See www.aio.tu-berlin.de/fileadmin/a3532/Kollegen/Bernhard/robert_roe_W_Onetworks.pdf.
29. "Bernhard was a prolific contributor to a host of international scientific organizations, many of which he founded or coordinated, including the European Network of Professors of Work and Organizational Psychology, the European Association of Work and Organizational Psychology, and New Technologies and Work (NeTWork)" (www.aio.tu-berlin.de/menue/about_us/in_memoriam_bernhard_wilpert/obituary/parameter/en).
30. See www.aio.tu-berlin.de/menue/about_us/in_memoriam_bernhard_wilpert/obituary/parameter/en.
31. Quantitative risk assessments.
32. Similar phenomena of transfer from one industrial activity to another also existed for companies or groups involved in a variety of domains. As an example, Rhône Poulenc, one of the first users of the ISRS in France started using it in its chemistry domain and it was then extended to others, like drugs. Likewise, within GE, "the ISRS started in plastic and then moved to GE power and then to GE medical" (consultant/industry 2, diverse industries, December 21, 2018).
33. From the HOF community.
34. However, interestingly, this potential bridge between the two safety science communities has limited effects. As mentioned earlier, it did not lead to a meeting of minds. The use of the Swiss cheese model by the reliability engineering community commonly diverted from its original intent that did not involve any quantification approach.
35. The Risø National Laboratory was a research laboratory created in the 1950s and focusing until the mid-1980s on providing stable and safe nuclear energy. Risø was among the active Nordic research centers that were pioneers in safety at that time.
36. The accident occurred on October 4, 1992.
37. ICAO Annexes booklet, p. 3 (www.icao.int/Documents/annexes_booklet.pdf).
38. See www.icao.int/safety/SafetyManagement/WebsiteDesignJuly2016/Initial%20Introduction%20of%20ICAO%20Safety%20Management%20SARPs.pdf.

39. Whether it is the case of aviation in general or just aviation safety would require further investigation.
40. The three presentations were the following (www.iapsam.org/psam-mtgs.html):
    - "Safety Substantiation Analysis for Airplane Systems," by B. Zilberman (Israel Aircraft Ind.)
    - "Reliability-Based Design of a Safety-Critical Automation System—A Case Study," by C. W Carroll (Boeing), W. Dunn (USC), L. Doty and M. Hulet (NASA), and M. V. Frank (Safety Factor Assoc.).
    - "System Safety Management in the UK Air Traffic Services," by R. Profit (Natl. Air Traffic Services).
41. See www.sandia.gov/LabNews/LN05-22-98/faa_story.html.
42. In order to ensure some independence from service provision, Eurocontrol has set up the Safety Regulation Commission (SRC), supported by the Safety Regulation Unit (SRU), which is responsible for the development of harmonized Eurocontrol safety regulatory objectives and requirements for the ATM system within the ECAC area.
43. This document is also known as the ESARR 3 (Eurocontrol Safety Regulatory Requirement).
44. Compared to other aviation activities.
45. The Single European Sky initiative was meant to address the issues related to the fragmentation of the airspace in Europe.
46. Source: http://dtitraining.blogspot.com/2013/01/faa-vs-transport-canada-sms-comparison.html.

REFERENCES

Aalders, M., & Wilthagen, T. (1997). Moving beyond command-and-control: Reflexivity in the regulation of occupational safety and health and the environment. *Law & Policy*, *19*(4), 415–443.

Almklov, P. G. (2018). Situated practice and safety as objects of management. In: *Beyond safety training* (pp. 59–72). Springer.

Amalberti, R. (1996). *La conduite de systèmes à risques*. Coll. Le Travail Humain. Presses Universitaires de France.

Beck, U. (1992a). From industrial society to the risk society: Questions of survival, social structure and ecological enlightenment. *Theory, Culture & Society*, *9*(1), 97–123.

Beck, U. (1992b). *Risk society: Towards a new modernity* (Vol. 17). Sage.

Bieder, C. (2021). Safety: A situated science. An exploration through the lens of Safety Management Systems. *Safety Science*, *135*, 105063.

Bieder, C., Le Bot, P., Desmares, E., Cara, F., & Bonnet, J.-L. (1998). *MERMOS: EDF's new advanced HRA method*. International Conference on Probabilistic Safety Assessment and Management. Springer Verlag London Limited.

Bird, F. E., Cecchi, F., Tilche, A., & Mata-Alvarez, J. (1974). *Management guide to loss control*. Institute Press.

Bourrier, M. (2005). The contribution of organizational design to safety. *European Management Journal*, *23*(1), 98–104.

Bourrier, M. (2017). Organisations et activités à risque: le grand découplage. In: Barbier, J.M. & Durand, M. (eds.), *Analyse des activités humaines. Perspective encyclopédique* (pp. 743–774). Presses Universitaires de France.

Bouwmans, I., Weijnen, M. P., & Gheorghe, A. (2006). Infrastructures at risk. In: *Critical infrastructures at risk* (pp. 19–36). Springer.

Cooper, S. E., Ramey-Smith, A. M., Wreathall, J., & Parry, G. W. (1996). *A technique for human error analysis (ATHEANA)*.

Cordel, F. (2019). *Gestion des risques et contrôle interne: de la conformité à l'analyse décisionnelle*. Vuibert.

David-Cooper, R. (2002). The transition to Safety Management Systems (SMS) in aviation: Is Canada deregulating flight safety. *Journal of Air Law and Commerce, 81*, 33.

Desroches, A., Baudrin, D., & Dadoun, M. (2009). *L'analyse préliminaire des risques: principes et pratiques*. Lavoisier.

Desroches, A., Leroy, A., & Vallée, F. (2003). *La gestion des risques*. Lavoisier.

Dien, Y., & Dechy, N. (2013), Les risques organisationnels des "organisations fragmentées." *Les entretiens du risques*.

DOT (Department of Transportation). (2008). *History of aviation safety oversight in the United States*. DOT/FAA/AR-08/39. US DOT.

EC (European Commission). (1983). *Framework programmes for Community research, development and demonstration activities and a first framework programme, 1984–1987*. European Union. https://cordis.europa.eu/programme/id/FP1-FRAMEWORK-1C

Eisner, H. S. (1976, July). Editorial. *Journal of Occupational Accidents, 1*(1).

Everdij, M. H. C., & Blom, H. A. P. (2016, August). *Safety methods database: Version 1.1*. Maintained by Netherlands Aerospace Centre NLR. www.nlr.nl/documents/flyers/SATdb.pdf

Gallagher, R. B. (1956). Risk management-new phase of cost control. *Harvard Business Review, 34*(5), 75–86.

Giddens, A. (1990). *The consequences of modernity*. Polity Press.

Gladwell, M. (2019). *Talking to strangers*. Little, Brown and Company.

Grimaldi, J. V. (2006 reprinted from November 1965). Management & industrial safety achievement: A practical approach and new philosophy for risk evaluation and control. *Professional Safety, 51*(2), 54.

Hale, A. R. (2003). Safety management in production. *Human Factors and Ergonomics in Manufacturing & Service Industries, 13*(3), 185–201.

Hale, A. R., & Glendon, A. I. (1987). *Individual behaviour in the control of danger*. Elsevier Science.

Hale, A. R., & Hovden, J. (1998). Management and culture: The third age of safety: A review of approaches to organizational aspects of safety, health and environment. *Occupational Injury: Risk, Prevention and Intervention*, 129–165.

Heinrich, H. W. (1931). *Industrial accident prevention: A scientific approach*. McGraw-Hill.

Hibou, B. (2012). *La bureaucratisation du monde à l'ère néolibérale* (p. 223). La Découverte, coll. "Cahiers libres." ISBN: 978-2-7071-7439-0

Hollnagel, E. (2008). The changing nature of risk. *Ergonomics Australia Journal, 22*(1–2), 33–46.

Hood, C. (1995). The "new public management" in the 1980s: Variations on a theme. *Accounting, Organizations and Society, 20*(2–3), 93–109.

Hovden, J. (1998). The ambiguity of contents and results in the Norwegian internal control of safety, health and environment reform. *Reliability Engineering & System Safety, 60*(2), 133–141.

Hovden, J., & Tinmannsvik, R. K. (1990). Internal control: A strategy for occupational safety and health: Experiences from Norway. *Journal of Occupational Accidents, 12*(1–3), 21–30.

Hutter, B. (2005). The attractions of risk-based regulation: Accounting for the emergence of risk ideas in regulation. *Centre for Analysis of Risk and Regulation Discussion Paper, 33*.

ICAO. (2002). *Line operations safety audit (LOSA)*. Document 9803. ICAO.

ICAO. (2016). *Annex 19: Safety management* (2nd ed.). ICAO.

Jasanoff, S., (2003). Technologies of humility: Citizen participation in governing science. *Minerva, 41*, 223–244. Kluwer Academic Publishers.

Kuhn, T. (1962). *The structure of scientific revolutions*. University of Chicago Press.

Lagadec, P. (1981). *Le risque technologique majeur: politique, risque et processus de développment*. Pergamon Press.

La Porte, T. R. (1996). High reliability organizations: Unlikely, demanding and at risk. *Journal of Contingencies and Crisis Management*, *4*(2), 60–71.

Le Coze, J.-C. (2012). *De l'investigation d'accident à l'évaluation de la sécurité industrielle: proposition d'un cadre interdisciplinaire (concepts, méthode, modèle)*. Gestion et management. Ecole Nationale Supérieure des Mines de Paris.

Lehtinen, J., & Ahola, T. (2010) Is performance measurement suitable for an extended enterprise? *International Journal of Operations & Production Management*, *30*(2), 181–204. https:// doi.org/10.1108/01443571011018707

Leveson, N. (2004). A new accident model for engineering safer systems. *Safety Science*, *42*(4), 237–270.

Lindøe, P. H. & Olsen, O. E. (2009). Conflicting goals and mixed roles in risk regulation: A case study of the Norwegian Petroleum Directorate. *Journal of Risk Research*, *12*(3–4), 427–441

Lupton, D. (Ed.). (1999). *Risk and sociocultural theory: New directions and perspectives*. Cambridge University Press.

Madsen, P. M. (2013). Perils and profits: A reexamination of the link between profitability and safety in US aviation. *Journal of Management*, *39*(3), 763–791.

Martínez-Lorente, A., Dewhurst, F., & Dale, B. G. (1998), Total quality management: Origins and evolution of the term. *The TQM Magazine*, *10*(5), 378–386.

Maurino, D. (2017). *Why SMS: An introduction and overview of safety management systems*. OECD. www.itf-oecd.org/sites/default/files/why-sms.pdf

Milch, V., & Laumann, K. (2016). Interorganizational complexity and organizational accident risk: A literature review. *Safety Science*, *82*, 9–17.

Mohaghegh, Z., & Mosleh, A. (2009). Incorporating organizational factors into probabilistic assessment of complex socio-technical systems: Principles and theoretical foundations. *Safety Science*, *47*, 1139–1158.

Perrow, C. (1984). *Normal accidents: Living with high risk technologies*. Basic Books.

Pidgeon, N. (2010). Systems thinking, culture of reliability and safety. *Civil Engineering and Environmental Systems*, *27*(3), 211–217.

Powell, T. C. (1995), Total quality management as competitive advantage: A review and empirical study. *Strategic Management Journal*, *16*(1), 15–37.

Power, M. (2000). The audit society—Second thoughts. *International Journal of Auditing*, *4*(1), 111–119.

Power, M. (2004). *The risk management of everything: Rethinking the politics of uncertainty*. Demos.

Power, M. (2007). *Organized uncertainty: Designing a world of risk management*. Oxford University Press.

Rasmussen, J. (1982). Human errors: A taxonomy for describing human malfunction in industrial installations. *Journal of Occupational Accidents*, *4*(2–4), 311–333.

Rasmussen, J. (1983). *Skill, rules and knowledge: Signals, signs, and symbols, and other distinctions in human performance models*. IEEE Transactions on Systems, Man and Cybernetics SMC-13 (3).

Rasmussen, J. (1985). Trends in human reliability analysis. *Ergonomics*, *28*(8), 1185–1195.

Rasmussen, J. (1997). Risk management in a dynamic society: A modelling problem. *Safety Science*, *27*(2), 183–213.

Rasmussen, J., & Batstone, R. (Eds.). (1989). *Why do complex organizational systems fail?* Summary Proceedings of a Cross Disciplinary Workshop on "Safety Control and Risk Management." World Bank.

Reason, J. (1988). Modelling the basic error tendencies of human operators. *Reliability Engineering & System Safety*, *22*(1–4), 137–153.

Reason, J. (1990). The contribution of latent human failures to the breakdown of complex systems. *Philosophical Transactions of the Royal Society of London. B, Biological Sciences, 327*(1241), 475–484.

Rochlin, G. I. (1997). *Trapped in the net: The unanticipated consequences of computerization.* Princeton University Press.

Schulman, P., Roe, E., Eeten, M. V., & Bruijne, M. D. (2004). High reliability and the management of critical infrastructures. *Journal of Contingencies and Crisis Management, 12*(1), 14–28.

Sills, D. L., Wolf, C. P., & Shelanski, V. B. (Eds). (1982). Accident at Three Mile Island: The human dimensions. Westview Press.

Spira, L. F., & Page, M. (2003). Risk management: The reinvention of internal control and the changing role of internal audit. *Accounting, Auditing & Accountability Journal, 16*(4), 640–661.

Stolzer, A. J., & Goglia, J. J. (2015). Safety management systems in aviation (2nd ed.). Ashgate.

Swain, A. D. (1964). *THERP* (No. SC-R-64–1338). Sandia Corp.

Swain, A. D. (1990). Human reliability analysis: Need, status, trends and limitations. *Reliability Engineering & System Safety, 29*(3), 301–313.

Top, W. N. (1991). *Safety & loss control and the International Safety Rating SystemTM (ISRS).* DNV. www.topves.nl/PDF/Safety%20Management%20and%20the%20ISRS.pdf

Turner, B. A. (1978). *Man-made disasters.* Wykeham Press.

Walters, G. A. (2000). *Aviation safety: Hearing before the subcommittee on aviation of the Committee on Commerce, Science, and Transportation United States Senate One Hundred Fifth Congress.* DIANE Publishing.

Weick, K., & Sutcliffe, K. (2015). *Managing the unexpected: Sustained performance in a complex world* (3rd ed.). Wiley.

Weijnen, M. P., & Bouwmans, I. (2006). Innovation in networked infrastructures. *International Journal of Critical Infrastructures, 2*(2/3), 121–132.

Wenger, E. (2011, October). *Communities of practice: A brief introduction.* STEP Leadership Workshop, University of Oregon.

Wynne, B. (1988). Unruly technology: Practical rules, impractical discourses and public understanding. *Social Studies of Science, 18*(1), 147–167.

6 Beyond the SMS
Toward More Contextualized Perspectives on Safety

Even though it is still presented as a step change in safety, at least in aviation, the SMS as currently practiced has a number of limitations as a safety enhancement approach (as developed in Chapter 4). Yet it fulfills numerous other functions that reach far beyond the safety purpose per se. These findings lead to a number of reflections and open several avenues to move forward. This chapter will present and discuss three proposals, ranging from an evolution of the SMS to a significant change in the way safety would be considered. The first proposal, an evolutionary one, builds on the existing SMS and reaches beyond some of the identified limitations while keeping the current foundations, especially the focus on risk management and the hierarchical control model. It mainly consists of suggesting several improvements to the risk management approach. The second proposal broadens the understanding of safety beyond the management of risks. It involves revisiting the foundations of safety management, especially acknowledging uncertainty and contingencies. Such a new framework calls for a number of changes at many different levels: conceptual, methodological, practical, and governance. Indeed, acknowledging that uncertainty is inevitable requires one to shift from an a priori hierarchical control model to other approaches allowing for living with such uncertainty. The third proposal consists of adopting a wider perspective—that is, to put safety back in context and to address it more appropriately. Considering safety as one among many interrelated stakes also calls for different concepts, methods, practices, and governance approaches. It especially involves a more transverse approach and an acknowledgment that all might not be easily apprehended through a set of indicators dedicated to each organizational stake. The conditions for such an evolution will be discussed in the light of an understanding of safety as coupled with a broad context.

6.1 MAKING THE SMS A MORE EFFICIENT SAFETY ENHANCEMENT APPROACH

Safety, as defined in the SMS, is "the state in which risks associated with aviation activities, related to, or in direct support of the operation of aircraft, are reduced and controlled to an acceptable level" (ICAO, 2016, pp. 1–2). With this definition of safety, the emphasis is put on risks and, as developed earlier, even more specifically on operational risks—that is, at the sharp end despite numerous evidence of other relevant contributions to safety or the absence of it. This section

DOI:10.1201/9781003307167-6

focuses on evolutionary ways forward, building on the current SMS philosophy and framework, considering risk management as the cornerstone. We propose several extensions of the current SMS risk management approach, not only in the type of the risks considered but also in the scope of the risk analysis and in the timeframe taken into account. Last, beyond the framework of the risk analysis, the last subsection involves proposals on the way of performing the risk analyses themselves.

6.1.1 A Broader View of Risk Management: Wider Scope, Time Frame, and Reach

Beyond Operations: A Wide Range of Actors and Aspects within the Organization and Beyond Contribute to Safety

Whereas the safety risk management pillar of the SMS puts the emphasis back on the sharp end and operations, there is a wealth of literature on safety developed since the late 1980s and 1990s that has illustrated the key role of organizational factors (Perrow, 1984; La Porte, 1996; Vaughan, 1996) summarized especially by Bourrier (2017). Numerous accident investigation reports refer to some elements considered as contributing factors that reach far beyond operational considerations, including organizational characteristics. It is the case, for example, of the recent merging of two airlines in the Dryden accident report (Moshansky, 1992) or of the wages policy in the Colgan Air accident report (NTSB, 2009). Likewise, regulatory oversight is sometimes argued to be another contributing factor, like the FAA's failure "in its oversight responsibilities to ensure the safety of the traveling public" in the case of Boeing 737 Max accidents (House Committee on Transportation and Infrastructure, 2020, p. 4). The idea that the system involved in risk management is much broader than the sharp end has been around for decades. The model proposed by Rasmussen back in 1997 already went from the sharp end to the government level. In other words, when trying to model safety or doing retrospective analysis, it is commonly acknowledged that safety is wider than operations. Yet none of these aspects are explored when focusing the risk analysis on operational aspects.

As suggested by Desroches et al. (2016) in the development and documentation of the "global risks analysis" method, a risk analysis can be made global by contemplating a whole range of possible hazards that reach far beyond operations to embrace, for example, political, financial, and cultural aspects (see illustration Figures 6.1 & 6.2). Even though less detailed on the operational aspects, accident scenarios starting from these kinds of hazards allow for identifying risks that are not currently addressed, although realistic. Examples of aspects contributing to safety mentioned by international trainees of an advanced master's in safety management in aviation at ENAC, based on their experience as aviation professionals, include stability of political environment, civil and military cooperation for airspace management, and independence of oversight. Extending the scope of possible hazards to phenomena beyond operations also leads to extending it to a broader timeframe and considering what may result from past decisions (including non-decisions) or measures as developed hereafter.

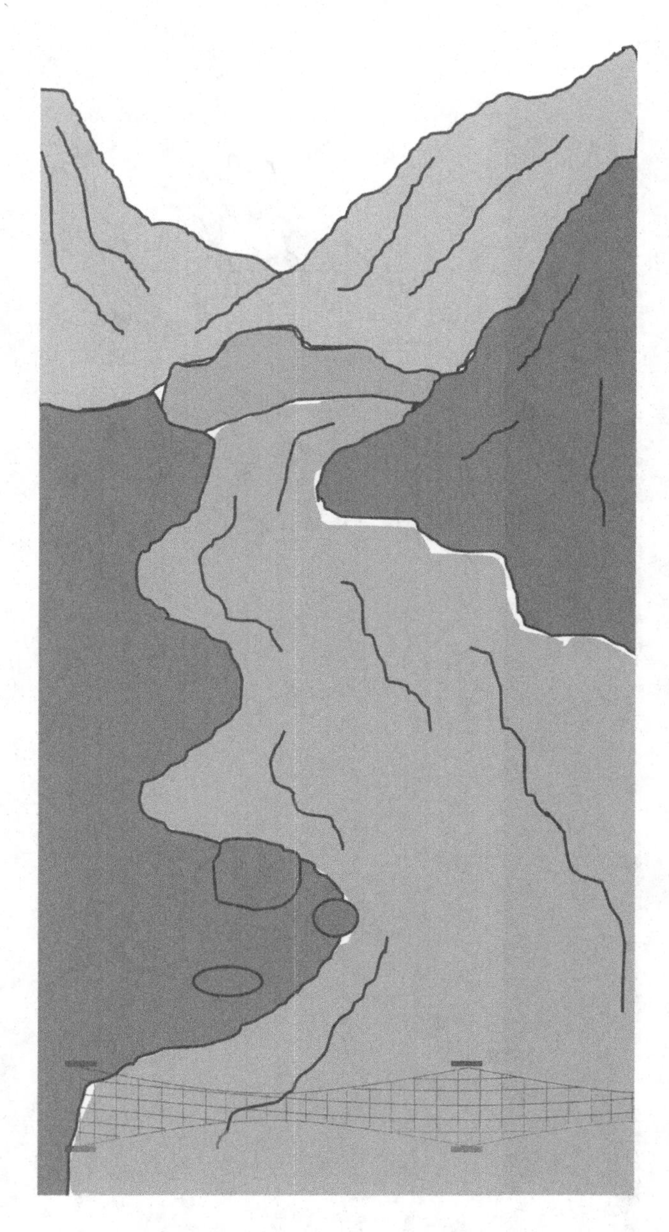

FIGURE 6.1 Focus on operational risks, here and now (visible from the surface).

Beyond Here and Now: A Wider Time Frame

Past Decisions and Actions Play a Role

Here again, the idea is not new. Going from the sharp end to the blunt end is often considered to go together with going from real-time to past-time, although it is not always the case.[1] For example, Le Coze, building on research by Vaughan and Snook,

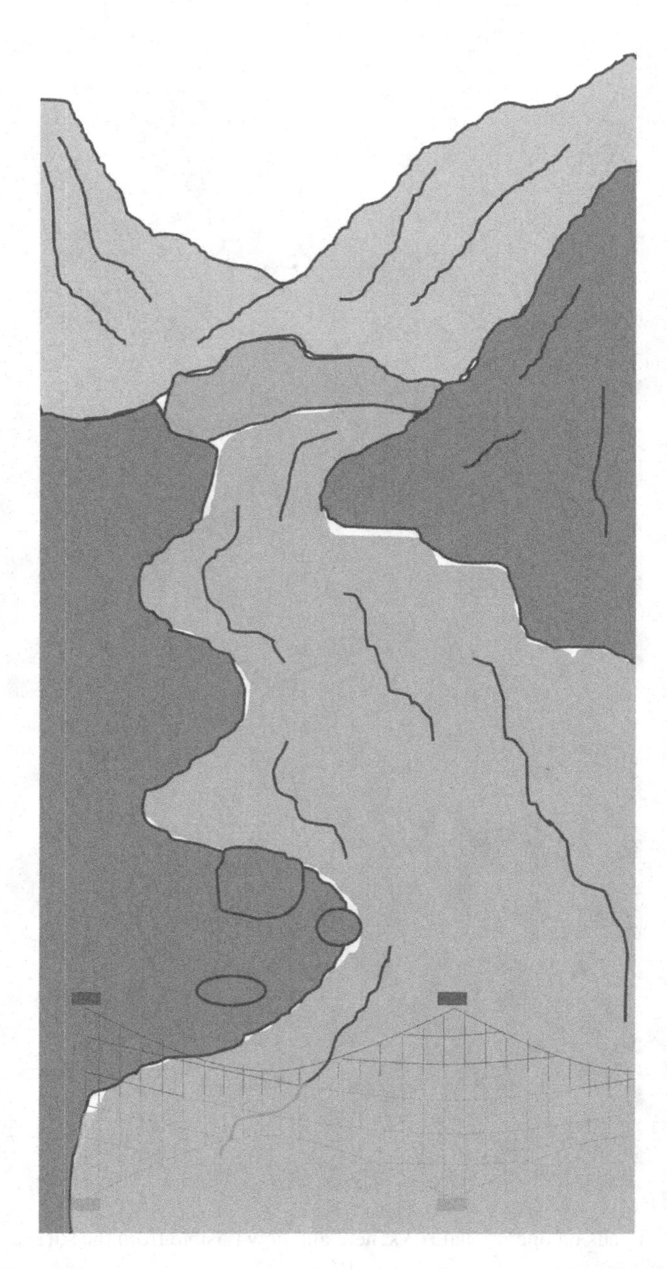

FIGURE 6.2 Wider range of risks, here and now (not all visible from the surface).

proposed in 2012 to consider all the changes, be they technical, organizational, personnel, or regulatory, over the past ten years, as they may induce changes in practices and thus in safety (Le Coze, 2012).[2] Sometimes, hazards emerge even more than ten years after decisions or changes. It is the case for example of pilots' manual skills eroding due to the decreasing training, including on-the-job training to manual

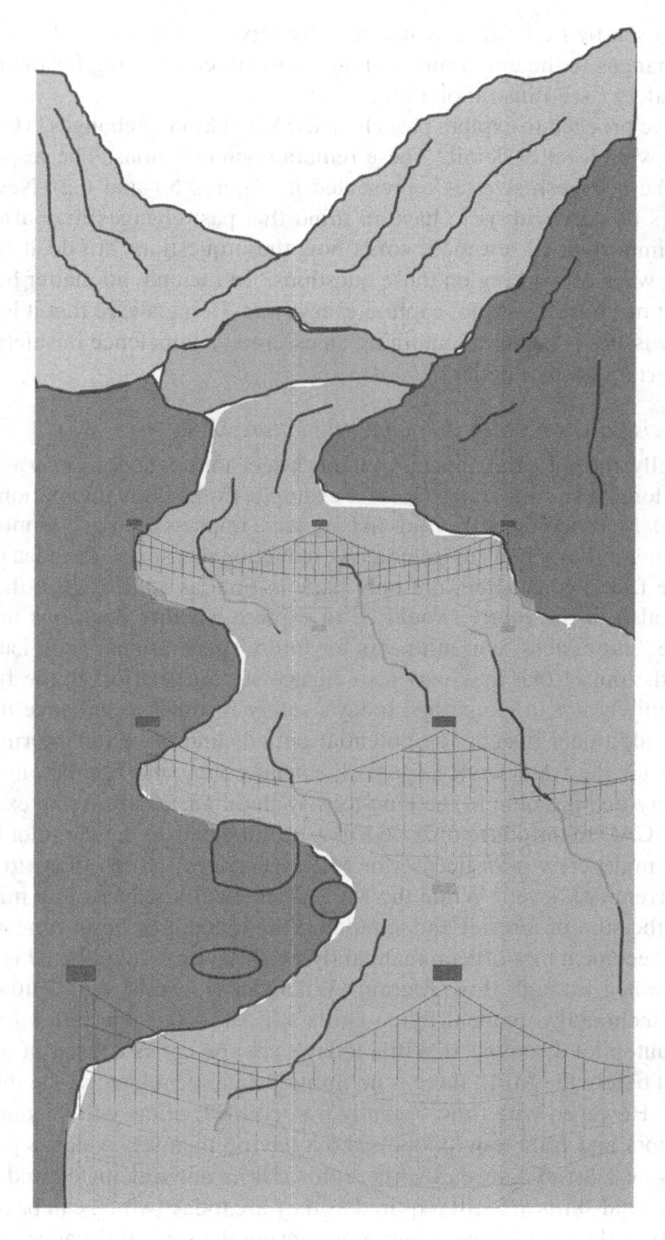

FIGURE 6.3 Wide range of risks, longer time frame (including upstream and not visible from the surface).

flying with automated systems being more reliable (since the late 1980s) and airlines' policies preventing pilots from manually flying above a certain altitude (commonly 1,000 feet or 500 feet) (Haslbeck & Hoermann, 2016; Casner et al., 2014). This latter example combines changes in technology but also in operational policies. Therefore, the list proposed by Le Coze may not be exhaustive, and changes of other natures, including changes in the environment, may be worth considering for their potential impact on safety (see illustration Figure 6.3).

How do we proceed to explore past changes? What kind of changes? How systematically? To what level of detail? These remain open questions. The responses will determine the net mesh size (as represented in Figures 6.1 and 6.2). Nevertheless, what matters to start with is to have in mind that past changes may affect safety. Equally as important (if not more so) is how these questions are dealt with in the sense of the ways of working on these questions.[3] In the end, no matter how performant the net mesh is, it will not capture everything. Being aware that it lets off part of the safety issues is key to maintaining an essential competence in safety—that is, keeping a certain level of doubt.

Today's Decisions and Actions Engage the Future of Safety

Symmetrically, the potential impacts of today's decisions or actions may affect safety in a mid- or long-term time frame (see, for example, the cockpit automation decisions in the 1980s). Methodologically, mid- to long-term impacts are not commonly envisaged. Neither are they when envisaging risk reduction measures. The idea of expanding the time frame considered, not only back in time as addressed in the previous section but also in the future, would be to explore whether decisions made today may impose "impossible" commitments for future "generations" or at least make it clear from the outset that they may also engage sustained effort in the future. The point is definitely not to jeopardize today's safety in order to enhance tomorrow's but to have additional insights on potential options and to better inform decisions and prepare for the framework (especially social policy and regulation) needed to support safety management in the long-term (Schulman, 2020). As an example, the decision by ICAO to introduce in 2006 a new training path to deliver pilot licenses—namely, the multi-crew pilot license, or MPL—engages future efforts to sustain at least the current risk level.[4] While the MPL dramatically reduces to a minimum of 240 hours (the sum of aircraft and simulator) the amount of flight time experience required to become a first officer, manual flying skills turn out to be a big challenge for pilots trained through this program (Wikander & Dahlström, 2016). Yet with the current technology, manual flying skills still constitute the last safety barrier, should the autopilot disconnect. While today's cockpit crews consist of at least one pilot trained before the MPL, there is no immediate issue related to the introduction of the MPL. However, with time, recently trained MPL pilots will become captains and instructors and train new pilots without having themselves developed manual flying skills (at least in a scenario where this risk is not anticipated and addressed per se). If manual skills are still required as they are today (which can be assumed at least for part of the fleet in operations, considering the several decades' life span of an aircraft), the introduction of the MPL, which involves very limited manual flying skills development, may lead to a risk in the future that may pose significant challenges to be reduced.

As for the extension of the time frame considered in the past, anticipating future potential safety impacts is not limited to decisions focused on safety, as developed earlier.

Beyond Organizational Boundaries: Envisaging Interfaces Collectively to Avoid Risk Transfer

Today's decisions and actions may be taken for good reasons within an organization and yet induce hazards or risks to other organizations. For example, an airline can decide to reduce its turnaround time (TAT) to increase its productivity, thus creating impossible conditions for ground handlers or de-icing companies to do their jobs properly and safely (both for their activity and for the safety of the flight to come).

Balancing an organization's interest against that of the overall industry is not natural. Yet it is a daily exercise, although somehow implicit, in interconnected infrastructures or systems of systems. Risks at the interface, however, are rarely addressed, for they are not clearly within the scope of one organization.[5] Yet creating the occasions for aviation organizations to interact with one another to exchange about the impact of their respective practices or decisions on others would be a way forward in reducing the risks of aviation operations by reaching beyond the individual organization's level. Although starting from a safety culture angle, the safety culture stack initiative (in the framework of the Future Sky Safety European program) experimented with such sessions where safety at the interfaces between specific airport stakeholders was discussed collectively with promising outcomes (Kirwan et al., 2018).[6] Putting different types of aviation organizations around the table is often revealing of their lack of knowledge of others' activities, objectives, and constraints. "We have been expanding knowledge and appreciation of how work is done on the other side . . . Just to learn how the other works can really improve understanding and alleviate tensions" (middle manager of a European aviation organization, May 2, 2017). At the interfaces where coordination is needed and requires efforts beyond a simple will to cooperate (McNamara, 2012), this lack of familiarity contributes to inducing hazards or risks to other organizations, usually involuntarily.

Addressing risks collectively involves new risk governance mechanisms, especially when other stakes beyond safety are involved. However, as mentioned earlier, the transfer of risks to another aviation organization is usually not performed knowingly or at least with a full understanding of the risks induced to this other organization. It is even more the case in interconnected infrastructures since all the organizations depend on one another to be able to operate.

6.1.2 Managing Risks: More in the Ways of Working Than in the Formal Processes, Methods, and Outcomes

Bringing together people from different backgrounds and perspectives, as suggested previously, to combine their experiences about real life and practices contributing to safety has often proved valuable for safety (Journé, 2018; Kirwan et al., 2018). It allows not only for highlighting the tensions faced by personnel in some situations but also for sharing them and discussing them with a broader understanding of the overall stakes, constraints, and challenges induced by the existing contradictions or conflicts between objectives (Kirwan et al., 2018).

Whereas risk management is commonly considered as a process consisting of the implementation of a set of methods or techniques leading to clear quantified outcomes and subsequent decisions, very little is said about the possible value of the ways of working as a contribution to safety.[7] Although the SMS requires setting up safety committees for the safety assurance part, they are presented as monitoring and decision-making bodies. The visible or auditable outcome of these safety committees (especially the minutes including the decisions) prevails over what may happen during these meetings. Likewise, nothing is said about the performance of the risk analysis itself or any other collective discussion and debate about tensions existing at all levels of the organization (although these tensions are likely to induce risks generic and "big" enough to be modeled in a risk analysis) or more contextual, dynamic, and detailed risks to be addressed in a different way.[8] As such, the performance of the risk analysis, provided the analysis is properly facilitated in groups, can be considered part of the learning mechanism and development of safety culture and greatly facilitates the mitigation of risks since they are apprehended and discussed from different viewpoints, and a collective and more realistic way forward can emerge and get buy-in from all stakeholders.[9] The involvement of different personnel, experts in their domain, and best witnesses of the reality of their activity is relevant not only in the initial phase of risk analysis but also in the identification and discussion of risk reduction measures. A parallel can be made here with the similar argument developed regarding the interfaces between organizations and the relevance of bringing representatives from interacting organizations to identify and manage risks at the interfaces to avoid risk transfer from one organization to another (see the previous section). The relevance of bringing different views together also applies within organizations between different divisions and hierarchical levels for similar reasons: better understanding the others' objectives and constraints and avoiding coming up with a solution that could jeopardize their activity.

Work involving staff from different divisions sharing different perspectives and experiences and collectively reflecting on the potential impact of some characteristics, measures, and decisions might be the most efficient safety enhancement. Weick and Sutcliffe (2015) already described the value of such practices to support the requisite variety needed to avoid simplification. It was also found to be current practice by managers from safe aviation organizations in Europe for decision-making when time permits, even though there is no underlying regulatory requirement or even internal process (Callari & Bieder, 2019). Beyond sharing perspectives and experiences, these exchanges can also be opportunities for highlighting irreducible tensions between safety and other stakes that can then be discussed to reach options that are more realistic. Eventually, the way of performing risk analysis or, more precisely, organizing the work and facilitating the methods is as important as, if not more so, the methods themselves.

This induces a change of perspective on risk management. Risk analysis cannot be the activity of a single person in his/her office, an expert in risk analysis methods. This also involves changes in the profile, training, and responsibility of the safety manager, as well as in the involvement of many other personnel in risk management. This assumes that gathering people from different divisions but also different hierarchical levels on a regular basis is possible. Conversely to safety committees, these groups are not necessarily established ones and gather people for their experience and even more for their capability to freely speak about what happens on a daily basis in reality. This freedom of speech, whatever the hierarchical level, requires at the organizational level, a framework or ideally a culture whereby hearing bad news is not bad news but rather a sign that one can have a clear view of real activity within the organization.

Such a change in perspective on the importance of the ways of working also involves an evolution of regulatory regimes that puts the emphasis on formal processes and outcomes more than on the ways of working. Even though the Australian CASA booklet is partly written in a way suggesting that the ways of working are important, falling back into a control philosophy and the development of indicators could be a way to seriously diminish the value of such approach. Some risks of regulating the approach to risk management in today's framework can easily be anticipated based on past experience. The choice of the participants and the content of the sessions would be secondary compared to the number of sessions. Providing risk analysis templates (or almost ready-to-use risk analyses as they are commonly used) to support organizations would not make sense. Indeed, the starting point then becomes the templates more than real-life experience, and the exercise becomes an adaptation exercise more than an actual reflection about the risks of the organization itself.

More generally, changing perspectives on risk management raises a number of questions that would be worth further exploring in particular: what conditions, especially organizational and managerial, but also regulatory and societal, are needed to make these ways of working possible?

6.2 REVISITING THE FOUNDATIONS OF THE SMS: ACKNOWLEDGING UNCERTAINTY AND ITS MANIFOLD IMPACTS

As discussed in Chapter 4, considering safety as the absence of unacceptable risks is a limited account of what contributes to ensuring a safe performance. Alternative views have been proposed—for example, the HRO theory—and sometimes opposed to the risk management one—for example, by Hollnagel with the distinction between Safety-I and Safety-II (Hollnagel, 2014). One of the main foundational differences beyond a change of focus from how things fail to how things work is the urge to account for what happens in the field and the associated acknowledgment of the complexity of work situations and the inevitable contingencies. However, acknowledging uncertainty and the impossibility to eliminate it is not obvious. It goes against the organizational, regulatory, and societal models and claims that have been prevailing over the past decades based on an *a priori* control through risk management. Furthermore, if controlling safety is not possible, living with the associated uncertainty (on top of that associated with the future) calls for other tools (e.g., concepts, methods, practices, regulatory regimes, legal systems) that are not yet available and articulated. This section goes back to a characterization of safety (here understood as reaching beyond risk management) and what it takes to maintain safe operations in the field. It then explores how today's context could be more favorable to wide recognition, at least intellectually, that uncertainty exists and cannot be totally eliminated. The multiple challenges of living with uncertainty are then reviewed and discussed in the light of today's context.

6.2.1 From Risk Management to Realities in the Field: Coping with Varying Complex and Sometimes Unanticipated Situations

Extending the scope of risk management, as suggested in the previous section, is not sufficient to overcome all the limitations developed earlier in Chapter 4. One of the major remaining issues is the focus on risks since it significantly limits the understanding of safety and, thus, of its management. Indeed, it is more and more commonly admitted not only by scholars but also by practitioners and regulators that safety cannot be strictly limited to maintaining risks at an acceptable level. Schulman, discussing the need for clarifying key concepts, summarizes the difference: "Safety is about assurance; risk is about loss" (Schulman, 2020, p. 74). Hollnagel, from a resilience engineering perspective, emphasizes that "safety is an active rather than a passive quality; safety is something that is produced, and which must continuously be produced, rather than something an organization simply has. Safety is therefore different from, and more complex than, the absence of risk" (Hollnagel, 2016, p. 73). In its summary and conclusions of a recent (2017) International Transport Forum roundtable on SMS, gathering international safety experts, the OECD stated that "safety must be seen as a state of effective functioning on all dimensions pertinent to the system's purpose and not merely the absence of risks, failures, and accidents" (ITF, 2018, p. 54).[10] Likewise, regulatory bodies also make a clear distinction between safety and the absence of unacceptable risk even though safety is still defined with respect to risks in the SMS requirements. As stated by the Safety Management International Collaboration Group (SMICG), a group of (almost 20 as of early 2020) aviation regulatory bodies, created in 2009 by the US FAA, the European EASA, and Transport Canada to promote a common understanding of safety management principles and requirements and facilitate their application across the international aviation community,

> Safety is more than the absence of risk; it requires specific systemic enablers of safety to be maintained at all times to cope with the known risks, to be well prepared to cope with those risks that are not yet known, and to address the natural "erosion" of risk controls over time.
>
> **(SMICG, 2013, p. 2)**

Although the latter statement keeps referring to risks, it points out a major distinction between what is known and what is not known (at least) yet. Precisely, one of the foundations of risk management is that it addresses the domain of the "knowable" as opposed to the domain of the unknowable (Desroches et al., 2003) (keeping in mind that these domains vary with time).[11]

> The elements of the knowable domain can be described qualitatively, but not necessarily their sequences. In this domain, the uncertainty is related to the complexity of processes and the incomplete knowledge about the parameters, variables and their value. However, the domain of the unknowable, by the very impossibility to observe, define or describe its elements remains out of the scope of risk management. This domain includes especially all that has not been imagined or is out of reach of observation because too small, too distant, belonging to the past or the future.
>
> **(Desroches et al., 2003, p. 25)**

Crises are examples of switching into this domain. Although risks are limited to the "knowable" domain, safety is not. Real work situations also consist of unanticipated or unthinkable phenomena (not only to the ones facing them but more generally). The "black swans," introduced by Taleb in 2001 (Taleb, 2001), or unknown unknowns, are part of this domain. Among the anticipated situations, Paté-Cornell (2012) distinguishes between "perfects storms," the ones belonging to the knowable domain—for example, the conjunction of individually known events or phenomena, the combination of which has never been thought out (aleatory uncertainty)—and "black swans," the phenomena no one could ever imagine, belonging to the unknown unknowns or the domain of the unknowable (epistemic uncertainty). Yet the author concludes that in practice, whether an unanticipated phenomenon is a perfect storm or a black swan does not make much difference.

Further, even though within the domain of the "knowable," everything can, in principle, be addressed as a risk, not all the risks can be anticipated and modeled. Perfect storms are examples of the ones that were missed. Indeed, the "required" level of detail, dynamics, evolution, and combination would make it impossible to represent all the risks at all levels. Risk management allows for identifying relatively big and stable issues that can be anticipated through rather generic models. But reality is not fully generic and involves a number of singularities. In addition, as discussed earlier, the accident scenarios identified through risk analysis are rather static. They cannot be exhaustive. In particular, they are defined at a level of granularity that overlooks a number of phenomena that can develop into an accident (especially emergences as they exist in complex systems). In other words, they do not provide sufficient insight to understand how to dynamically keep an organization safe.[12] Safety also involves all these real-time activities at all hierarchical levels that allow for coping with real situations in all their complexity (not accessible to models), all the contingencies where the a priori arrangements are not sufficient to cope with the situation.

Some of the characteristics of organizations good at managing the daily changing situation requirements or the unexpected or uncertainty, or the so-called resilient, have been analyzed and documented in the literature, especially HRO, resilience, and managing the unexpected (see, for example, La Porte, 1996; Weick & Sutcliffe, 2001, 2011, 2015; Schulman et al., 2004; Hollnagel et al., 2006; Grote, 2009). They describe organizational settings and practices that escape risk modeling and are not fully compatible with a hierarchical control model. Schulman et al. (2004) thoroughly analyze the practices of high-reliability professionals and distinguish between four different modes of performance depending on the system instability and the option variety (just-in-time, just-in-case, just-for-now, just-this-way), as well as the adaptability of professionals and their capability to navigate from one to the other being the source of the system's reliability (combination of production and safety). The authors also underline the ability of high-reliability professionals to navigate between sources of knowledge from representational to experiential and well as between scopes from system-wide (all cases) to specific events (a single case). In other words, what is needed at an operational level to ensure safety (and even beyond safety, a performance that balances production and safety) is not one approach better than another but the appropriate articulation of approaches suited to the situation and the purpose.

However, with the SMS approach based on a hierarchical control model, any performance mode but just-this-way is considered a deviation and discouraged.[13] There is no room for adaptative performance modes. Yet a thorough field analysis highlights that it is the way high-reliability professionals

> actively and consistently protect our infrastructures against disturbances, failures, and mistakes that could bring them down, including errors at policy and supervisory levels. They work, often heroically, against odds that many in the public, academia, and government can hardly appreciate.
>
> **(Roe & Schulman, 2008, p. 13)**

When one starts to appreciate the odds, one also needs to recognize the adaptations and real-time adjustments that are needed to cope with them. Acknowledging the necessity to adapt operational performance modes to varying and sometimes unpredictable conditions calls for other safety management "modes" beyond SMS. To build on high-reliability professionals, different performance modes, and their ability to navigate from one to the other, what might be needed are different safety management modes and the capability at all levels to navigate from one to the other. However, it all starts with acknowledging uncertainty and the impossibility of anticipating everything and manage safety through a sole risk management approach.

Roe and Schulman (2008) underline that the odds against which high-reliability professionals work are hardly appreciated by many in the public, academia, and government (see the previous quote). Organizations, beyond operational professionals, could be added to the list of those hardly appreciating the contingencies professionals face in their daily activities, beyond the public, academia, and government. This lack of appreciation may result from different reasons, among which are a true belief that reality works as described by the book or a kind of denial of the existence of contingencies and unanticipated situations by a lack of appropriate tools (e.g., concepts, regulatory regimes, legal systems) to deal with them.

6.2.2 Acknowledging Uncertainty: Increasing Evidence at All Levels but Remaining Confusion

Extensive scientific literature is available on the inevitable contingencies faced by first-line operators and the necessary real-time adjustments they make to maintain a safe performance. Could anyone still ignore that, despite a vast proceduralization, all situations are not anticipated and cannot be? Scientific literature may not reach organizational management or regulatory bodies or the public. It may not even reach academics from other safety science communities. However, at an organizational level, the phenomena at play that were highlighted were organizational blindness or simplification rather than pure ignorance. What is more often highlighted within high-risk organizations is the "necessary hypocrisy" requiring from employees a "smart" compliance with procedures by lack of another reference model.

Real situations and practices may be less easily accessible to authorities, especially in regulatory regimes where inspections are planned and tight in resources, and even more so when the oversight is limited to the SMS.

Real practices and especially the need for adjustments are not publicized outside high-risk organizations. Therefore, the public may ignore that the current strategies of safety management based on a priori risk control are not self-sufficient due to the inevitable contingencies that operational people have to cope with. Nevertheless, today's context may foster an overall evolution of the public's attitude toward uncertainty and unanticipated situations. Uncertainty and the impossibility to control everything a priori are made more and more tangible to the civil society through an increasing number of unexpected and unanticipated phenomena known to the public worldwide, at least in two areas. Security with some mediatized security events, like 9/11, the Paris attacks, the Germanwings accident in 2015, or a series of other, more-recent attacks. These events have highlighted (and unfortunately continue to do so) that uncertainty does exist (besides epistemic uncertainty) and cannot be eliminated. Climate change is another area where spectacular events (e.g., fires, droughts, thunderstorms) unexpectedly hit some regions. Although the nature of the threats can be anticipated to a certain extent, their magnitude, timing, or complex ripple effects remain unexpected. These repeated events may contribute to making the public realize and acknowledge uncertainty, or at least stand back from the idea that everything can be controlled a priori.

Yet some confusion remains as to what uncertainty is and, thus, how to live with it. A first step could be to clarify the various understandings and characterizations of uncertainty. Indeed, risk management itself claims to address uncertainty even though it is mainly through the probability of occurrence of anticipated phenomena. Rowe (1994) suggests a different view by distinguishing between four classes of uncertainty: temporal (both in future and past states), structural (due to complexity), metrical (related to measurement), and translational (related to the explanation of uncertain results and the underlying interpretative framework). According to the author, all four classes exist in all situations but not with the same importance. Dupuy and Grinbaum (2005) distinguish between epistemic uncertainty (i.e., related to the state of knowledge available) and intrinsic uncertainty (i.e., due to the random nature of the event under consideration). Further research work might be useful beyond the already existing characterizations of uncertainty involving the public to better apprehend today's attitudes toward the various uncertainties.

6.2.3 Living with Uncertainty: Revisiting the Main Challenges in Today's Context

Acknowledging that a priori control is insufficient and that we need to safely live with uncertainty requires more than an intellectual exercise if it is to translate into practice. As mentioned earlier, a reason for denying uncertainty might be the lack of appropriate tools to cope with it more than a genuine belief that it can be eliminated. Some of the challenges to considering safety as more than risk management have already been identified and discussed. Nevertheless, it might be worth reconsidering them in the light of today's context to discuss the potential remaining obstacles or, conversely, enabling conditions to overcome them.

Safety Concepts and Methods: From Competition to Articulation

One important and lasting challenge is conceptual and methodological. As developed in Chapter 5, safety science turns out to be a rather competitive research field where two communities, the reliability and safety engineering community and the HOF and social science community, continue to progress in parallel, hardly ever exchanging with one another. Within these macro communities, several schools of thought coexist, sometimes with very limited discussions, if any, between them. Concepts and methods tend to be developed one against the other with very limited articulation effort, if any (Schulman, 2020). Yet if a wider understanding of safety accounting for the complexity and possible unpredictability of work situations and several safety management modes are to be envisaged, several safety approaches might need to be combined. Some authors calling for the reconciliation or integration of different approaches to safety analysis have already highlighted the multiple challenges that it involves, especially methodological, practical, and political (Schulman, 2020; Le Coze, 2019). Some proposals are nevertheless put forward. For example, elaborating on the difference between safety management and risk management Schulman (2020) suggests the development of metrics to detect the "full" implementation of SMS—that is, reaching beyond the formal and structural aspects of the SMS. "The integration of SMS metrics with physical and engineering analyses can lead to a more powerful socio-technical understanding of complex systems, their operation and their risks" (Schulman, 2020, p. 77). The author also advocates the need for a higher-resolution safety management framework, considering different scales and extended timeframes to support this wider understanding. This could be one step forward to making safety science communities get closer. However, pushing the integration too far, including the development of metrics for aspects that are not so easily measurable, could lead to losing the value and complementarity of the various safety approaches. Without aiming for integration, further reflection of the articulation of the different views at different scales and time frames could be a way forward. Yet today's context may not offer significantly more incentives than in the past for academics from diverse disciplines to work together. At least the first step could be to try and map the scope, validity, and relevance domains of safety approaches to start making them complementary. It could highlight that they are not systematically competing but sometimes just addressing different scopes (e.g., known, knowable, unknown).

Governance Challenges: From Standardization to More Contextualized and Inclusive Approaches

However, moving away from the SMS, an easily auditable control model, is not just a matter of having good concepts, methods, and understanding. It also relies on a wider context (especially industrial, regulatory, societal) and requires transitioning from understanding to acting. Envisaging different safety management modes adapted to the situation also challenges one of the historical pillars of safety—namely, standardization and risk governance: standardization of practices at first-line operators' level through procedures, at the organizational level through processes, at the national level through regulatory requirements, or even at the industry level through industry standards. Although the SMS was meant to move in this direction of more customized approaches leaving leeway to organizations, the SMS as practiced largely comes back to implementing standards (e.g., organizational structure, risk analysis content, codes of management practices, indicators).

In recent years, the limitations of standardized approaches to safety have been underlined by a number of actors (Olsen et al., 2019). The challenge of standardization of safety was, for example, a conclusion from the experts from the International Transportation Forum (ITF, 2018): "Even within a single domain—aviation or maritime, for instance—there will be such a variety of operators and business conditions that standards become impossible" (ITF, 2018, p. 53). In fact, challenging safety standards naturally flow from the acknowledgment that safety is situated, contextualized, and coupled with a local and historical environment. The argument commonly advanced to support standards in aviation is the fact that aviation is an international activity by essence and that safety must be the same everywhere. This argument, although understandable and valid, leads to a shortcut by considering that how safety is "ensured" should also be the same everywhere, regardless of the organizational, local, and historical context.[14] Similar dilemmas regarding the standardization of safety approaches are faced by large organizations operating internationally with no obvious good solution (Guillaume et al., 2018). Adapting safety approaches to local contexts would challenge the dominant model of hierarchical risk control, not to mention the practical consequences. Yet remaining with centralized approaches keeps away a wealth of local knowledge that would help make models more adapted and efficient in a given context. It thereby prevents reaching beyond major inefficiencies of some safety measures based on top-down standardized approaches (Bourrier & Bieder, 2018). The case of the management by the World Health Organization (WHO) of the Ebola outbreak in West Africa between 2014 and 2016, analyzed by Bastide (2018), provides compelling illustrations of the ineptitude of safety management approaches thought and decided independently from the context in which they were meant to be implemented. Contextualizing safety approaches by involving local representatives is an interesting way forward in this respect, but also more generally.

Indeed, as mentioned earlier, the scientific production abounds on the unexpected uncertainties and the description of how individuals and organizations manage to ensure safe operations in the face of these unanticipated situations beyond hierarchical control. Still, there is a persistent gap between theory and practice, description and action. This knowledge has not translated into practice (Bourrier, 2017). Regulatory requirements remain focused on control and transparency, pushing for easily auditable and measurable evidence. There is little if any discussion or debate with the civil society (originally calling for control and transparency) regarding how organizations actually operate and the necessary adjustments and flexibility to ensure safety. Exposing this reality within the public arena, beyond the official "clean" model where everything is anticipated, and an a priori safe response developed that operators or organizations just have to comply with, is still considered indecent (Gilbert, 2018).

The civil society, although highlighted as a key influencer through the societal expectations in the development of safety approaches, remains a distant stakeholder and is hardly ever involved directly in safety-related discussions and debates. Even when there are mandatory calls for involving local populations, the consultations are organized in such a way that it makes it hard for the civil society to actually participate in public debates (Kamaté, 2018). Safety performance indicators, what they actually indicate, and what they also contribute to hiding are never debated, either within an industry or with external stakeholders. More generally, there is very limited involvement of the civil society in safety governance or the development of safety approaches. The impermeable frontier between safety governance and the civil society induces several challenges to progress in safety management. In particular, it prevents objective discussions and debates about today's societal expectations with respect to safety and the approach to uncertainty and to excellence in safety versus the balancing of multiple stakes. In this respect, it prevents reaching beyond the current illusion of control despite its known limitations.

As suggested by Jasanoff (2003), involving the civil society in areas where uncertainties are great, such as safety, is even more important. This turns out to be a recent trend in research (see, for example, the call from the European Commission: "Science with and for society").[15] Furthermore, according to Matyjasik and Guenoun (2019), the new public management could come to an end with new forms of governance possibly emerging to replace it. Among the alternative forms they envisage is the new public governance, where the civil society plays an important role. These trends could create favorable conditions to move toward more contextualized and inclusive approaches that, if implemented in the safety domain, could help reach beyond current obstacles and induce significant safety progress.

6.3 DIMMING THE SPOTLIGHT ON SAFETY TO PUT IT BACK IN CONTEXT: A CONDITION TO BETTER APPREHEND IT

Whatever the refinement of the definition of safety and the sophistication of the methods used to apprehend it, managing safety does not take place in a void but rather in a complex environment where other stakes also are at play. Whether considered as an obstacle to production, for example, or a constraint or a cost, proposed safety enhancement measures are not always supported. Initiatives to improve safety culture, aimed in theory at making sure safety is given appropriate importance, are themselves proposed as safety enhancement measures that may well not be supported or at least not in the spirit of the former concept. In short, identifying safety enhancement measures does not make it all. The path to implementing them may be long, difficult, and just a dead end.

This section comes back to the interrelations between safety and other stakes at different levels, whether individual, organizational, or governmental. Building on existing developments, it then reflects upon the implications of considering safety in competition with other stakes and highlights some limitations to such an approach. An alternative view is proposed where safety is rather considered in coopetition (cooperative competition) with other stakes. The implications and challenges of such a proposal are then discussed.

6.3.1 Safety: One Stake among Others, Neither Isolated nor Stand-Alone

Although "safety first" is often used as a motto in high-risk organizations, observing and analyzing how safety is addressed in practice leads to qualifying this statement and opening to other stakes. Indeed, whether at an individual, organizational, institutional, or even political level, safety or its management is never the unique objective or *raison d'être*. The work presented in Section 5.3 already provided some illustrations of the multiple interrelated objectives, partly conflicting, that are at stake even when safety is supposed to be the core purpose. For example, the authorities' motivations to move toward the SMS involved liability and resource issues as much as a safety enhancement. For industrials, improving the overall performance of the organization (in an increasingly competitive environment) was part of the motivation. As for the insurance companies, cost reduction was an important driver. Examples of intertwined interests, of which safety is one among others, at all levels abound.

Baram and Lindoe (2018), for example, illustrate the trade-off made by authorities in the case of the Chevron refinery in Richmond (California) between its societal value as key actor of the economy and its poor safety management:

> Federal and state regulators, knowing that the refinery's operations are of considerable importance to the national and state economies, dutifully take enforcement actions to punish non-compliance but tailor them to avoid impacting operations to the extent that their societal value would be impaired.
>
> **(Baram & Lindoe, 2018, p. 73)**

Closer to aviation, the *Preliminary Investigative Findings*, published by the House Committee on Transportation and Infrastructure in March 2020, on the recent case of the Boeing 737 MAX, states,

> There was tremendous financial pressure on Boeing and subsequently the 737 MAX program to compete with Airbus' A320neo aircraft. Among other things, this pressure resulted in extensive efforts to cut costs, maintain the 737 MAX program schedule, and not slow down the 737 MAX production line. The Committee's investigation has identified several instances where the desire to meet these goals and expectations jeopardized the safety of the flying public.
>
> **(House Committee on Transportation and Infrastructure, 2020, p. 3)**

This example illustrates the trade-offs between finance and safety made at the highest level of the industrial company.

At the middle managers' level, balancing multiple stakes when making decisions is part of the daily activity. Callari and Bieder (2019) analyzed the contribution of managers to safety and investigated their associated practices. The authors highlight that safety is never considered in isolation, as a dimension they manage per se but rather as one stake, a key one, but still interplaying with many others. Balancing safety with other stakes sometimes turns out to be challenging depending on the organizational culture: "some middle or top-managers in the organization are focused on other 'core business' indicators (e.g. costs, productivity)" (Bieder & Callari, 2020). When suitable to the situation requirements, the middle managers with a safety-oriented mindset even tend to make decisions collectively and involve people from different divisions and with different backgrounds to take into account a variety of aspects, safety being one of them but not the only one (Callari & Bieder, 2019).

At the individual level, the management of simultaneous objectives, including safety, has been extensively analyzed from different perspectives. Fucks and Dien (2013), for example, highlight the trade-offs made by some first-line operators between the safety of operations and their liability. It could translate into hiding behind procedures even when knowing that they are not fully efficient in some situations safety-wise or in not doing anything outside of a procedure, although it would make sense safety-wise. Roe and Schulman (2008) look at individual performance differently, in a more integrative manner. They highlighted the various performance modes depending on the system volatility and the options variety of high-reliability professionals to continue to deliver the service in the face of changing demand while ensuring safety. In that way, they do not oppose safety to production upfront or refer to trade-offs. Interestingly, in the HRO theory, safety is from the outset addressed as intertwined with production rather than in isolation as it is commonly considered.

Eventually, in practice, whatever the level, safety is never an isolated stake that would exist in a kind of void and could be addressed per se. Yet it is often the way it is presented, understood, and considered in many industries and safety approaches or theories.[16]

6.3.2 Safety as Competing with Other Stakes: The Trade-Off Perspective (A Zero-Sum Game)

Some authors discussed the limitations of such a way to consider safety and envisaged it in a broader framework. Amalberti, in 2013, elaborated on the necessary trade-offs to navigate/manage safety (Amalberti, 2013), even discussing safety as an "art of trade-off" (Amalberti, 2017), a bargaining game almost. Considering that safety is one risk among others, "an assented risk to maximize other benefits" (ibid., p. 26), the author discusses the necessary trade-offs made, especially at the top-management level, between safety and other stakes and the risk emerging from the absence of plan B in case the ideal measures to manage safety are not decided. According to the author, preparing a plan B—that is, a degraded way of managing safety anyway (measures to keep safety acceptable as opposed to optimal)—would reduce the risk of seeing the ideal safety plan turned down in the trade-offs involving not only safety but also the other major stakes of organizations.[17] In other words, considering safety as one stake among others would lead to envisaging not only ideal safety measures (if safety were not "constrained" by any other aspect) but also less ambitious (yet sufficient) safety measures (i.e., a plan B) and sometimes even bottom-line safety measures since organizations have other objectives beyond managing safety. However, in this perspective, safety is still considered as an identifiable activity as such (to which a budget can be allocated and dedicated) and may still appear as a constraint or enemy to other stakes or "real business" (what generates revenue rather than exclusively costing resources).

6.3.3 Safety as Part of the Same Boat as Other Stakes: The Conjoint Perspective

At the other extreme is the view suggested by Hudson and Stephens (2000), where "safety *makes* money, allowing operations to take place that would be impossible without an adequate safety regime" (p. 1). Safety would then be seen as a necessary condition but almost in the sense of a blackmailing game, "no safety, no business," still somehow disconnected from the "real business."

Another way to look at safety in a broader context is to think about it as fully integrated into the business. It is neither considered in isolation nor opposed to the other stakes. They are simply on the same boat, all necessary to make it move forward. Rather than a mere competition on resources, coexisting and partly conflicting stakes can be seen from a different angle where synergies are also possible. As put forward by one of the interviewees, a number of organizations turned to the ISRS and continued to use it because they were convinced it would not only improve their safety performance but rather the company's overall performance. Beyond the view that safety is a cost that needs to be balanced with other costs (and investments), safety measures can also pay off in other areas beyond safety. The HRO theory, for example, does not oppose safety and production but rather considers them jointly and collaboratively. Safety is not just what makes the business possible, or a threat to the business, or a constraint of the business; it is the business together with the other objectives.

With this view, any decision, not only technological or operational but also in other domains, like finance or HR, for example, is likely to affect safety (as well as any other aspect) positively, negatively, or both. A number of illustrations were provided by past research on middle managers' contributions to safety (Callari & Bieder, 2019; Bieder & Callari, 2020). For example, the authors highlight the impact of HR policies on safety. The existence of a "talent" career path in the exemplified European aviation organization makes it hard to reward and sometimes to keep employees long enough in a job or a domain to develop the necessary competencies and experience to stand back and have a safety-critical eye on what they do. Likewise, the authors highlight that employees who are driven by key performance indicators (their professional yearly objectives) focused on cost and production may tend to disregard safety to a certain extent. Another organizational choice, the multiplication of processes (not only for safety), is also identified as affecting safety (as well as other stakes) for employees no longer tend to think by themselves and make sense of what they do. They do things because they are written, without trying to understand. Conversely, the authors also illustrate the synergies in middle managers' practices between safety and other organizational objectives. "People from my team inform me if there is a problem but there is no difference between safety and other aspects. It is a matter of developing a trust climate. I proceed in a similar way for all aspects" (middle manager, European aviation organization, March 10, 2017).

Such a perspective on safety does not prevent "independent" dedicated safety experts or studies from analyzing in-depth the safety-related aspects. Nevertheless, the safety-related recommendations consider that safety does not exist in a void and can be discussed and debated jointly with the experts of the other organizational objectives in a power game where collaboration and concertation are more valued than competition. In a way, looking at the various organizational stakes as interconnected infrastructures or from an inter-organizational collaboration perspective (except that it would rather be intra-organizational) could open new avenues to explore, especially in terms of governance. A parallel could be drawn between the two situations in the ambiguity of the relationships between various stakes in the first case and various organizations in the second, even though they cannot be considered fully similar. "Interorganizational collaboration is intriguing, in particular, because of its paradoxical nature, combining competition and cooperation, and autonomy and interdependence" (Rodriguez et al., 2007, pp. 150–151). Likewise, the concept of coopetition, although defined in a different context to characterize inter-organizational relationship, could provide an interesting lens to analyze the ambiguous interplay between safety and other organizational objectives: "coopetition is a paradoxical relationship between two or more actors simultaneously involved in cooperative and competitive interactions, regardless of whether their relationship is horizontal or vertical" (Bengtsson & Kock, 2014, p. 182). Yet in the coopetition framework, the cooperation part is always serving the competition part.

Other perspectives could be explored to move away from a competition idea that involves an opposition. The evolution that happened with the concept of safety culture could be inspiring. Making sure that safety is given appropriate importance within an organization was the starting point of safety culture in the late 1980s. Initially identified as a useful concept and way forward to ensure that safety would be given sufficiently considered, safety culture was later criticized for it conveys the idea that there would be a distinct culture for safety as opposed to an overall organizational culture. Through the very name "safety culture," safety is isolated from the rest. Much literature exists on the topic, with a clear school of thought getting back to organizational culture and considering that safety is part of it (Gilbert et al., 2018). There is no such thing as a safety culture that would exist in parallel with the organizational culture or in competition with the production culture, for example. It is the overall culture that makes sense, and this culture balances all its objectives, not one by one but together. This approach does not deny the value of analyzing the importance given to safety in organizations but considers that it does not result from a dedicated isolated culture or dedicated measures. Along similar lines, regarding safety management in general, analyzing safety in its milieu or analyzing the milieu with a special look at safety could be an interesting avenue to reach beyond a perspective isolating safety from the other organizational stakes. It would involve considering that all practices, not only those identified as safety ones, which may have an impact on safety as opposed to considering that there are such things as safety practices and that they are what ensure safety. To build on the work presented earlier, even the adoption of the SMS as a regulatory requirement cannot be considered a safety decision per se but rather illustrates the interplay between the various organizational objectives, where safety is balanced with resources constraints, control, and liability requests among others at the authorities' level.

Today, to a certain extent, safety as an activity (or dedicated activities) is the wood for the trees. The existence of a dedicated safety organization, of clearly defined processes and responsibilities, or of safety performance indicators is considered sufficient to operate safely even though they mainly come down to accountant's activities or "counting safety" activities. Yet they responded at least decades ago to societal expectations for control and liability.

The proposed evolution of perspective can be illustrated as shown in Figure 6.4. In the second case, so-called safety activities stop to stand out to the benefit of a

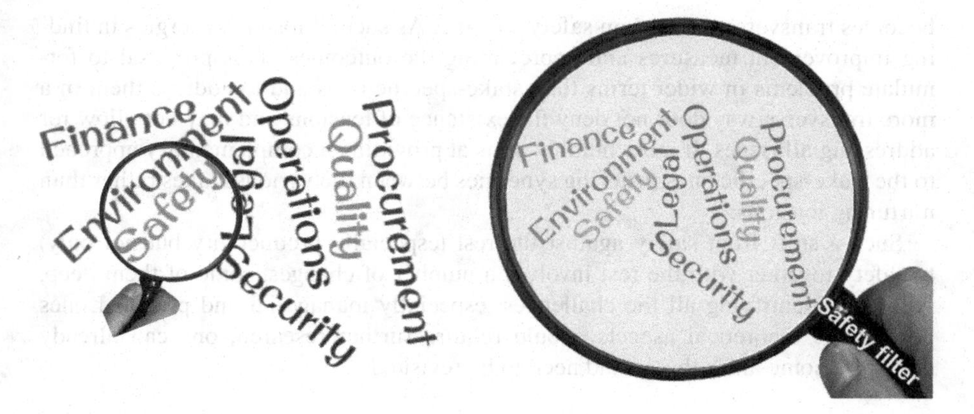

FIGURE 6.4 Isolating safety versus looking at an organization from a safety viewpoint.

global analysis of how safe the organization's functioning in a context where safety is not the only aspect to be managed and interplays with many other stakes. Yet this global analysis can be performed through a safety lens/filter—that is, from a safety viewpoint.

By making safety disappear from the center stage where it was artificially placed and adopting a safety lens to look at organizations, safety is given back a more realistic and promising place.

6.3.4 Fostering Synergies versus Nurturing Tensions: Proposal and Challenges

Even though considering safety in its milieu as opposed to a stand-alone activity is already a start, it does not say much about the lens through which one can apprehend safety. As developed earlier, focusing on the competition between safety and other objectives tends to isolate safety, as well as all the other objectives, and view them as opponents in a zero-sum game. An alternative perspective would be to analyze what safety and the other stakes share and exchange, what jointly benefits safety and the rest, as much as the tensions that exist between these partly conflicting stakes. In such an approach, a so-called safety measure would not be considered to benefit safety alone. More generally, some improvement measures would not be specifically labeled as safety, production, or quality. For example, as advanced by Hudson (2001) regarding safety culture and reinforced by one of the interviewees, the increase in trust in managers that can result from safety-culture-related initiatives does not impact safety alone. Fostering the synergies between the various stakes, be it at an individual, organizational, or societal level, may start with stating problems differently, more upstream than their stake-specific declination. Highlighting the issue of trust in managers is a good illustration of standing back from a safety-specific problem of insufficient reporting (even though it is not the only upstream condition to foster reporting). Stated as such, the problem becomes transverse rather than safety specific. As such, it fosters synergies in finding improvement measures and appreciating the outcomes. This proposal to formulate problems in wider terms than stake-specific ones and to address them in a more transverse way does not deny the existence of tensions and may not allow for addressing all issues. It more humbly aims at providing a complementary approach to the stake-specific ones fostering synergies between the various stakes rather than nurturing tensions.

Such a shift from safety against the rest (especially productivity but not only) to safety together with the rest involves a number of changes, some of them deep. Although identifying all the challenges, especially managerial and practical ones beyond the theoretical aspects, would require further research, one can already anticipate some areas that would need to be revisited.

Definition and Measurement of Performance

Considering that all the objectives or stakes are interplaying with one another (and not only competing) makes the definition of performance and its indicators most often per objective largely insufficient. Not only do these stake-related indicators nurture the idea of a competition between them, thus of a disconnection between them, but also they disregard all the interrelations, whether they are rather synergies or tensions or a mix of both. In other words, through what they follow and highlight, they contribute to hiding a wealth of significant aspects. More generally, considering that the various objectives interplay may challenge the need for or illusion to control and "count" the performance even more. Indeed, as illustrated earlier with the trust in managers example, some of the synergetic aspects may be hard to measure and quantify.

Organizational and Governmental Structures

Organizational structures themselves could also be challenged by this more systemic lens, for they generally tend to correspond to (induce or follow) the split between objectives through the delineation of finance, production, HR, quality, safety, and other departments or divisions. The same applies to governmental structures or at least ways of working.

Governance Models

Revisiting the definition of performance and how to apprehend it requires the involvement of a variety of actors, both within organizations but also outside. No easy answer can be proposed that would reflect the various views of the various stakeholders. Debates around the question of performance and measurement are needed to address such complex questions and, at the same time, contribute to a wide acknowledgment of their complexity. Besides, acknowledging that safety is not a specific activity resulting from dedicated practices but shares some aspects with other stakes also challenges the possibility of standardizing it. As such, it also challenges the current governance models where, for example, international standards and recommended practices are the common basis of safety regulation all around the world.

The examples only provide a flavor of what it would take to change perspectives and try to foster synergies among the different stakes. Ultimately, adopting a systemic perspective on organizations where safety cannot be isolated but considered as interrelating with other stakes has manifold impacts at many different levels. It challenges concepts, theories, but also disciplinary splits, as well as organizational structures, tools, practices, and governance structures and models. It opens many avenues for further research, including how to conduct systemic research.

6.4 CONCLUSION: CONTEXTUALIZATION, INCLUSION, AND HUMILITY AS COMMON DENOMINATORS WHATEVER THE AMBITION

One can envisage several ways forward to reach beyond the limitations of the SMS as a safety enhancement approach. From an incremental improvement by extending the scope of risk management and revisiting the ways of working to much deeper changes, such as the extension of the definition and scope of safety, or even to the contextualization of safety within a more complex set of interrelated stakes. Although the reach and magnitude of the implications vary from one option to another, they all call for more contextualization, inclusion, and humility.

For the first proposal, the improvement of the risk management approach, taking into account a broader organizational scope, a historical perspective and organization, and field-specific inputs, supports the contextualization of the risk analysis. The involvement of employees with different perspectives and experiences in the risk management approach makes it not only more contextualized but also more inclusive. Additionally, the various inputs to and debates around the risk analysis or appropriate risk mitigation measures highlight the variety of perspectives and inevitable drawbacks of each solution. In that respect, it incites humility.

For the second proposal, suggesting a broader definition of safety, contextualization is one of the main drivers since its purpose is to reach beyond generic risk models. Furthermore, it calls for inclusion in many respects: the involvement of a variety of actors within the organization, but also outside with a necessary evolution of governance modes, the combination of different safety management methods often coming from different disciplines. Accounting for the variety and complexity of operational situations, or realizing the methodological gap to apprehend it, illustrates the limitations of safety management in this framework and thus calls for humility as well.

The third proposal is built around even more contextualization since it suggests putting safety back into a wider context of interrelated stakes. Beyond high-risk organizations themselves, it also suggests considering the overall societal views on safety in a context where the perception of other aspects, such as climate change or security, are evolving. As mentioned earlier, the challenges associated with such a proposal are huge and concern a wide range of aspects and actors. Addressing them in line with the philosophy of the proposal naturally involves inclusion at many different levels: inclusion of several disciplines at the academic level, inclusion of different stakes representatives, and inclusion of industry, society, and government to come up with new adapted governance structures and models. The complexity, especially the interrelations between the various stakes and the various spheres (e.g., political, societal, industrial), highlighted by the proposed perspective on safety calls even more for humility.

NOTES

1. For example, an airline CEO calling for no go-arounds would be an action taken at the blunt end with possible very short-term effects, as soon as a crew is in a situation that would require a go-around (and this can be anytime for many different reasons). Conversely, a deviation recurrently performed at the blunt end may only manifest itself and contribute to an accident much later. Likewise, some maintenance errors at the sharp end only have an impact months or years later (e.g., foreign objects sometimes cause damage to aircraft systems long after having been left behind during a maintenance operation).
2. Although the Swiss cheese model developed by Reason also considered all sorts of levels and their impact on safety, it mainly focused on failures at all these levels.
3. Ways of working as a key aspect of safety enhancement will be discussed later, in Section 6.1.2.
4. Presented as both a response to the anticipated pilot shortage considering the commercial air transport market growth forecast and a focus on the key competences with the evolution of pilots' jobs over time.
5. Whether the initial risk (before any risk reduction measure is taken) or the residual risk (the remaining risk after risk reduction measures are implemented).
6. Future Sky Safety is a program funded by the European Commission under the H2020 framework. It started in 2015 and consisted of five safety projects, one of which was concerned with "resolving the organizational accident." As part of this project, one work package was dedicated to safety culture and the extension of the concept to several interacting aviation organizations.
7. Research on ways of working to improve risk management was conducted in the field of complex projects risk management. It highlighted the benefits of agile project management for risk management, especially when combined with complex system theory allowing for capturing interdependencies between risks (Marle, 2020).
8. Other ways forward beyond risk management will be addressed in the following section.
9. This outcome dismisses the resources argument claiming that such analysis requires more resources than a quick and individual one.
10. From regulatory agencies, academia, research institutes, consultants, and investigation bodies.
11. As developed in Section 4.2.1.
12. Keeping an organization safe is an "improper" statement used for the sake of readability. In reality, it is as safe as reasonably possible with all the discussions that it involves.
13. Section 6.3 will further expand on the relevance of considering other stakes as well together with safety.
14. In reality, there are some nuances introduced in SMS regulation, but mainly related to the complexity of operators.
15. See https://ec.europa.eu/programmes/horizon2020/en/h2020-section/science-and-society.
16. As mentioned previously, the HRO theory is an interesting exception. Even though it was sometimes challenged as a safety approach, for it focuses on reliability rather than safety, reliability includes the balancing of production and safety in this framework.
17. One could see an extension of the notion of sufficient performance as opposed to optimal performance advanced two decades earlier at the level of first-line operators (Amalberti, 1996).

REFERENCES

Amalberti, R. (1996). *La conduite de systèmes à risques*. Coll. Le Travail Humain. Presses Universitaires de France.

Amalberti, R. (2013). *Piloter la sécurité: théorie et pratiques sur les compromis et les arbitrages nécessaires*. Springer.

Amalberti, R. (2017). *La sécurité industrielle est-elle un art du compromis ? Audit, risques & contrôle, 12*.

Baram, M., & Lindoe, P. (2018). Risk communication between companies and local stakeholders for improving accident prevention and emergency response. In: *Risk communication for the future* (pp. 61–77). Springer.

Bastide, L. (2018). Crisis communication during the Ebola outbreak in West Africa: The paradoxes of decontextualized contextualization. In: *Risk communication for the future* (pp. 95–108). Springer.

Bengtsson, M., & Kock, S. (2014). Coopetition—Quo vadis? Past accomplishments and future challenges. *Industrial Marketing Management*, *43*(2), 180–188.

Bieder, C., & Callari, T. C. (2020). Individual and environmental dimensions influencing the middle managers' contribution to safety: the emergence of a "safety-related universe." *Safety Science*, *132*, 104946.

Bourrier, M. (2017). Organisations et activités à risque: le grand découplage. In: Barbier, J.-M. & Durand, M. (eds.), *Analyse des activités humaines. Perspective encyclopédique* (pp. 743–774). Presses Universitaires de France.

Bourrier, M., & Bieder, C. (Eds.). (2018). *Risk communication for the future: Towards smart risk governance and safety management*. Springer International Publishing.

Callari, T. C., Bieder, C., & Kirwan, B. (2019). What is it like for a middle manager to take safety into account? Practices and challenges. *Safety Science*, *113*, 19–29.

Casner, S. M., Geven, R. W., Recker, M. P., & Schooler, J. W. (2014). The retention of manual flying skills in the automated cockpit. *Human Factors*, *56*(8), 1506–1516.

Desroches, A., Aguini, N., Dadoun, M., & Delmotte, S. (2016). *Analyse globale des risques: principes et pratiques*. Lavoisier.

Desroches, A., Leroy, A., & Vallée, F. (2003). *La gestion des risques. Principes et pratiques*. Lavoisier.

Dupuy, J. P., & Grinbaum, A. (2005). Living with uncertainty: From the precautionary principle to the methodology of ongoing normative assessment. *Comptes Rendus Geoscience*, *337*(4), 457–474.

Fucks, I., & Dien, Y. (2013). "No rule, no use"? The effects of over-proceduralization. In: *Trapping safety into rules how desirable or avoidable is proceduralization* (pp. 27–39). CRC Press.

Gilbert, C. (2018). Safety: A matter for "professionals"? In: *Beyond safety training* (pp. 1–9). Springer.

Gilbert, C., Journé, B., Laroche, H., & Bieder, C. (Eds.). (2018). *Safety cultures, safety models: Taking stock and moving forward*. Springer International Publishing.

Grote, G. (2009). *Management of uncertainty: Theory and application in the design of systems and organizations*. Springer Science & Business Media.

Guillaume, O., Herchin, N., Neveu, C., & Noël, P. (2018). An industrial view on safety culture and safety models. In: *Safety cultures, safety models* (pp. 1–13). Springer.

Haslbeck, A., & Hoermann, H. J. (2016). Flying the needles: Flight deck automation erodes fine-motor flying skills among airline pilots. *Human Factors*, *58*(4), 533–545.

Hollnagel, E. (2014). *Safety-I and safety-II*. Ashgate.

Hollnagel, E. (2016). Safety management: Looking back or looking forward. In: *Resilience engineering perspectives* (Vol. 1, pp. 77–92). CRC Press.

Hollnagel, E., Woods, D. D., & Leveson, N. (2006). *Resilience engineering: Concepts and precepts*. CRC Press.

The House Committee of Transportation & Infrastructure. (2020). *The Boeing 737 MAX aircraft: Costs, consequences, and lessons from its design, development, and certification: Preliminary investigative findings*. https://transportation.house.gov/imo/media/doc/TI%

20Preliminary%20Investigative%20Findings%20Boeing%20737%20MAX%20March%202020.pdf

Hudson, P. T. W. (2001). Safety management and safety culture: The long, hard and winding road. *Occupational Health and Safety Management Systems, 2001*, 3–32.

Hudson, P. T. W., & Stephens, D. (2000, January). *Cost and benefit in HSE: A model for calculation of cost-benefit using incident potential.* SPE International Conference on Health, Safety and Environment in Oil and Gas Exploration and Production. Society of Petroleum Engineers.

ICAO. (2016). *Annex 19: Safety management* (2nd ed.). ICAO.

ITF. (2018). *Safety Management Systems*. International Transport Forum Roundtable Report, OECD Publishing.

Jasanoff, S. (2003). Technologies of humility: Citizen participation in governing science. *Minerva, 41*(223–244). Kluwer Academic Publishers.

Journé, B. (2018). How to deal with the contradictions of safety professional development? In: *Beyond safety training* (pp. 103–110). Springer.

Kamaté, C. (2018). Public participation in the debate on industrial risk in France: A success story? In *Risk communication for the future* (pp. 17–30). Springer.

Kirwan, B., Reader, T., & Parand, A. (2018, July). The safety culture stack: The next evolution of safety culture? *Safety and Reliability, 38*(3), 200–217. Taylor & Francis.

La Porte, T. R. (1996). High reliability organizations: Unlikely, demanding and at risk. *Journal of Contingencies and Crisis Management, 4*(2), 60–71.

Le Coze, J.-C. (2012). *De l'investigation d'accident à l'évaluation de la sécurité industrielle: proposition d'un cadre interdisciplinaire (concepts, méthode, modèle).* Gestion et management. Ecole Nationale Supérieure des Mines de Paris.

Le Coze, J.-C. (2019). Resilience, reliability, safety: Multilevel research challenges. In: *Exploring resilience* (pp. 7–13). Springer.

Marle, F. (2020). An assistance to project risk management based on complex systems theory and agile project management. *Complexity*. doi:10.1155/2020/3739129

Matyjasik, N., & Guenoun, M. (2019). *En finir avec le New Public Management*. Nouvelle édition [en ligne]. Institut de la gestion publique et du développement économique (généré le 18 juin 2019). Disponible sur Internet: http://books.openedition.org/igpde/5756

McNamara, M. (2012). Starting to untangle the web of cooperation, coordination, and collaboration: A framework for public managers. *International Journal of Public Administration, 35*(6), 389–401.

Moshansky, V. (1992). *Commission of inquiry into the air Ontario crash at Dryden, Ontario.* Final Report. Minister of Supply and Services.

NTSB. (2009), *Accident report, NTSB/AAR-10/01, PB2010–910401.* www.ntsb.gov/investigations/AccidentReports/Reports/AAR1001.pdf

Olsen, O. E., Juhl, K. V., Lindoe, P. H., & Engen, O. A. (2019). *Standardization and risk governance: A multi-disciplinary approach* (p. 306). Taylor & Francis.

Paté-Cornell, E. (2012). On "Black Swans" and "Perfect Sorms": Risk analysis and management when statistics are not enough. *Risk Analysis, 32*, 1823–1833.

Perrow, C. (1984). *Normal accidents: Living with high risk technologies*. Basic Books.

Rodriguez, C., Langley, A., Béland, F., & Denis, J. L. (2007). Governance, power, and mandated collaboration in an interorganizational network. *Administration & Society, 39*(2), 150–193.

Roe, E., & Schulman, P. R. (2008). *High reliability management: Operating on the edge.* Stanford Business Books.

Rowe, W. D. (1994). Understanding uncertainty. *Risk Analysis, 14*(5), 743–750.

Safety Management International Collaboration Group. (2013). Measuring safety performance guidelines for service providers. www.skybrary.aero/bookshelf/books/2395.pdf

Schulman, P. R. (2020). Integrating organizational and management variables in the analysis of safety and risk. In: *Human and organisational factors* (pp. 71–81). Springer.
Schulman, P. R., Roe, E., Eeten, M. V., & Bruijne, M. D. (2004). High reliability and the management of critical infrastructures. *Journal of Contingencies and Crisis Management, 12*(1), 14–28.
Taleb, N. N. (2001). *Fooled by randomness: The hidden role of chance in life and in the markets*. Random House.
Vaughan, D. (1996). *The Challenger launch decision: Risky technology, culture, and deviance at NASA*. University of Chicago Press.
Weick, K. E., & Sutcliffe, K. M. (2001). *Managing the unexpected.* Jossey-Bass.
Weick, K. E., & Sutcliffe, K. M. (2011). *Managing the unexpected: Resilient performance in an age of uncertainty* (Vol. 8). John Wiley & Sons.
Weick, K. E., & Sutcliffe, K. M. (2015). *Managing the unexpected: Sustained performance in a complex world* (3rd ed.). Wiley.
Wikander, R., & Dahlström, N. (2016). *The Multi-Crew Pilot License part II: The MPL data-capturing the experience*. Lund University School of Aviation.

7 Conclusion

The objective of this work was to investigate how the SMS was adopted in aviation in the mid-2010s with the promise to be a step change in the management of safety and a way to reduce the occurrences of accidents. This question was all the more sensible to me that I had had the opportunity to see the SMS implemented in several aviation organizations and hear feedback from practitioners, but I wanted to have a more scientific approach to it. Building on an in-depth analysis of the SMS in aviation and of the genesis of SMSs, the author suggests several ways forward to enhance safety that reaches beyond the current limitations of the SMS, from a limited evolution to a more revolutionary one challenging the very fact to focus on safety to address it.

This concluding chapter starts with an overview of the investigation work presented in this book. The second part comes back to the proposals derived from these insights to enhance safety management. The third part includes a discussion on the interest in conducting a multidisciplinary research. In the light of these elements, the fourth part provides a synthetic response to the initial question of the SMS as a step change in aviation safety.

7.1 INSIGHTS FROM DIFFERENT VERSIONS OF THE SAFETY MANAGEMENT SYSTEM

Having observed the implementation of the SMS in several aviation organizations for the past decade, either as an insider employee, an external expert, or a researcher, and noticing some significant limitations safety-wise, confirmed by feedback from practitioners and scientific papers, was a strange, if not bitter, feeling. Starting from there, understanding why aviation considered in the mid-2010s (and still does) the SMS as a promising way forward to enhance safety became a question worth investigating scientifically. Doing so required a multifaceted approach. First, it involved getting back to what the SMS is and what promises it conveys in aviation. Second, based on this in-depth analysis of the SMS in aviation, it meant characterizing its limitations. Third, this apparent paradox led me to dig deeper to find out why aviation adopted the SMS in the mid-2010s and in the first place to investigate the emergence and spreading of the SMS decades earlier.

The in-depth analysis of the aviation SMS relied on document analysis and literature review in addition to an auto-ethnographic approach. Having worked in aviation safety for the past 20 years, I have had the opportunity to witness the emergence of the SMS in aviation and observe as an insider the questions and reactions raised by its implementation progressively becoming mandatory by regulation. This specific positioning as part of the aviation safety community also gave me privileged access to SMS documentation that is normally considered proprietary and rarely shared publicly. Nevertheless, referring to this written material in a research work is not possible. Therefore, the documents used to perform an in-depth analysis of what the SMS is in aviation (at least for its written traces) were regulations, guidance material, and resources kits put together by CAAs from different countries. The selection of document sources was meant to reflect various approaches to the SMS, of philosophies and interpretations, and thus, of guidance that are not anecdotal and used in a very limited number of aviation organizations but rather widely used worldwide. Therefore, the main documents used were that developed by (1) the UK CAA, known in Europe and beyond for its very detailed guidance; (2) the US FAA since many countries around the world use either the US FAA or the European EASA documents as a reference for their own regulatory frameworks; and (3) the Australian CASA for the singularity of its SMS philosophy, Australia having a significant aviation background and a remarkable safety record. In addition to these documents, the literature on SMS in general and SMS in aviation more specifically was reviewed to benefit from existing research findings. This in-depth analysis of the SMS and, more specifically, the aviation SMS allowed for highlighting several of its key characteristics. First, it is promoted as a step change in aviation safety by allowing for addressing safety in a more proactive way, especially through a risk analysis. Furthermore, it involves switching from compliance to a performance-based approach whereby each aviation organization is given a leeway to define the improvement measures most appropriate to its own characteristics. It is also meant getting the management's attention to safety by turning safety into a business function. Finally, the SMS should allow for improving safety performance and continuously reducing the number of aviation accidents and incidents. In practice, the SMS regulatory requirement relies on four pillars—namely, (1) safety policy and objectives, (2) safety risk management, (3) safety assurance, and (4) Safety promotion. Safety risk management advanced as the cornerstone of the SMS focuses on operational risks. As for safety performance, it is addressed through indicators that may vary dramatically from one organization to another, from accident/incident numbers or administrative processes to less operational aspects that may have an impact on safety. More generally, although the four pillars of the SMS (and the 12 elements they consist of) are common to all aviation organizations, they are translated and interpreted in diverse ways in guidance material and resources kits and may lead to very different implementations from one organization to another. At one extreme, implementing the SMS requirements can be turned into a mere compliance exercise if not a cut-and-paste exercise consisting in changing the name and logo and making minimal adjustments to the resources developed by CAAs. At the other extreme, what is fostered is the self-reflection of aviation organizations, the resources being focused on supporting this reflection (e.g., methodologies, ways of working) rather than providing ready-to-use templates and analyses.

The critical review of the aviation SMS was based on the outcome of the in-depth analysis and on other sources, especially scientific literature review, as well as gray literature and exchanges with practitioners. It allowed for highlighting mixed perceptions of the SMS, including very skeptical ones. Some practitioners consider that the SMS brings nothing new except for the packaging, or that it is the ultimate proceduralization of safety, or even that it has led to freezing the reflection on safety for a number of years. Beyond these perceptions, three types of limitations were highlighted. First, conceptual shortcuts and assumptions. In particular, it considers that a good SMS enhances safety performance, whereas no link between the two has been established scientifically. Further, safety assurance is addressed by means of safety performance indicators with no guarantee that these indicators are related to the actual safety performance. Additionally, safety is defined exclusively as a risk management approach. It thereby disregards all the domain of unanticipated situations or the uncertainty not restricted to the calculation of probabilities. The risk management approach promoted by the SMS also involves some limitations—for example, by focusing on operational risks—thus disregarding other types of hazards that may have a more indirect or delayed impact on operations. Last, the managerial dimension of safety is mainly seen through accounting lenses, leaving apart aspects that are not so easy to regulate or quantify, such as leadership, for example. Besides these conceptual limitations, the SMS also involves methodological limitations. Even though there is no method imposed as a support to risk analysis and risk management, the bow-tie approach turns out to be the most used in SMS. Yet although it is acknowledged as a useful method to communicate about risks, it is a limited proxy to actually analyze and manage risks. Among the main shortcomings are the wording of consequences, making it impossible to assess the severity and the associated probability (the two components of risk according to the risk management definition); the diverse intrinsic criticality of top events; the linear modeling of accident scenarios; and the limited number of accident scenarios modeled. Last, some practical limitations of the SMS were also highlighted. Considering the qualitative and quantitative resources needed to implement the SMS according to its initial intention—that is, in a reflexive manner—leaving leeway to organizations to determine the corrective measures best suited to their characteristics, in many cases the actual implementation looks rather like a cut-and-paste exercise with little customization.

Since the in-depth analysis of the SMS confirmed its limitations as a safety approach, another analysis was performed from a different angle to understand why the SMS was adopted in aviation after all. Further insights were searched through a socio-historical perspective of the genesis of the SMS and its spreading across high-risk industries, including its late landing in aviation. This analysis relied on a historical approach, combining written sources and oral history. It involved the search for and analysis of historical documents and archives on safety management. These documents included, for example, the archives from ICAO; the programs of safety-related conferences, especially in the 1980s and 1990s; and a report on an international workshop held by the World Bank on safety management in the late 1980s. Besides, 18 old-timers who have been in the safety domain for decades were interviewed. They were chosen for they played a role in the development or implementation of safety management approaches, including at the time when the SMS emerged. Further, the interviewees came from seven different countries and represented a range of safety management stakeholders and high-risk industries: five interviewees were from regulatory bodies, three from high-risk industries, four from academia, two from consulting companies on safety management, and four had hybrid profiles including two or more of these experiences either sequentially or simultaneously.

The socio-historical analysis of the genesis of the SMS provided another explanation for the emergence and spreading of the SMS.

The intellectual context around safety and the way to manage it was flourishing with new ideas and concepts in the late 1970s, 1980s, and 1990s. Although the historical safety science community, focusing on safety and reliability engineering, continued to develop, refine, and share methods deriving from engineering, another safety science community historically focused on human factors and human performance developed to better account for the organizational aspects of safety. Social scientists especially described and characterized high-reliability organizations or, more generally, the organizational aspects of safety, showing that what contributed to the safety of an organization's performance was more complex and subtle than a command-and-control approach. Likewise, they highlighted that accidents could result from other phenomena than the failure of such an approach, such as the normalization of deviance or the organizational blindness to any phenomenon not corresponding to its beliefs.

The analysis also highlighted a combination of motivations reaching far beyond safety from all safety stakeholders to move toward a new safety approach. These motivations included, for example, cost reduction for insurance companies, the improvement of their overall performance for industrials, protecting their liability and increasing their efficiency for some regulators, coping with the limitation of resources for others, and growing defiance toward public services (thus, regulators) from the civil society. All this happened in a context in which high-risk industries were evolving at a fast pace not only in terms of technology but also of competition, industrial organization, and development. Furthermore, total quality management was becoming a trend, making management systems part of the common landscape in an overall society claiming for more transparency and efficiency of public services. It was the advent of the new public management, calling for the use of private sector methods trusted as efficient in public services.

Ideas on safety management mainly circulated through communities with very few bridges between them, relying essentially on personal trajectories and career paths, funding initiatives fostering the mix of participants, or post-accidental international collaboration. The communities were either industry-specific communities of practice, communities of users (especially that of the ISRS), or scientific communities with two parallel branches, one consisting of safety and reliability engineers massively coming from the industry with some regulators, another one gathering, mainly academics from human and social sciences and some individuals from the industry. However, the relative influence of the two safety science communities could not compare with one another. The engineering community was far better introduced among industrials and regulators.

Ultimately, despite the new developments in safety science, the SMS, an approach largely inspired by quality management systems and in line with the safety and reliability engineering safety science community, made its way through. It became the new accepted standard responding to the many individual motivations of the various stakeholders and the global societal and industrial trends.

The aviation domain was not yet part of this big evolution toward the SMS, not at that time. First of all, it was a very closed world, with hardly any exchange with other industrial fields. Most exchanges took place among aviation actors. Historically, safety was mainly ensured by the intrinsic safety of technology. Design of technology completed by an extreme proceduralization and selection and training of aviation professionals was the recipe for safety. The main push for a change in safety management approaches came from regulators. In the US, the flattening of the accident rate curve over the 1970s to the 1990s and the anticipated traffic growth alerted the need to improve safety and keep public confidence. Besides, the investigation of the ValuJet accident (1996) pointed out the lack of anticipation by the FAA of the impacts of rapid growth and massive outsourcing of maintenance. In Canada, the Air Ontario accident in 1989, that is two years after airline deregulation, highlighted organizational factors that contributed to the accident. It led Transport Canada to develop a safety management approach integrating organizational aspects. In Europe, the push came instead from air traffic control with the privatization or corporatization of ANSPs, starting in 1996 with the UK National Air Traffic Services. This new configuration urged to regulate safety aspects independently from economic aspects. Under the strong influence of the UK that had already adopted and promoted the SMS in many other high-risk industries, Eurocontrol developed an SMS regulatory requirement. Considering the international and highly regulated nature of the aviation field, the wide adoption of the SMS in aviation went through the publication by the ICAO of a recommendation to implement an SMS applicable to all aviation organizations. How the idea made its way through to the ICAO combines a change of the head of the Air Navigation Bureau and ways of working of ICAO on safety and the influence from Transport Canada and Eurocontrol that were ahead with the SMS.

7.2 THE PROPOSED WAYS FORWARD

Contrasting the different versions of the SMS leads to proposing three avenues for improvement of safety management.

The first one, the least challenging considering the current safety management framework, keeps the basics of the SMS, especially the definition of safety as the management of risks. However, it suggests both (1) extending the scope, time frame, and reach of the risk analysis and (2) revisiting the ways of working to perform risk analyses. The extension of the scope consists of considering other aspects than those directly related to operations. Organizational characteristics may also significantly impact the safety of operations, as illustrated by many accident investigation reports, although not necessarily directly or in real time. Accounting for these remote or non-immediate influences also calls for extending the time frame considered in risk analysis. Some events, decisions, or phenomena that took place at a given moment in time can induce a risk much later (e.g., the automation of aircraft in the 1980s led to a progressive decrease in pilots' manual handling skills that is now considered a safety issue). Likewise, some events, decisions, or phenomena occurring today may turn out to have long-term consequences on safety. Besides, the proposed improvements of the risk analysis include addressing the risks at the interfaces between organizations. This involves joint work between organizations interacting with one another. Beyond addressing this currently gray area at the interface, bringing organizations together allows for improvement of the mutual understanding of their respective activities and avoids the inadvertent generation of hazards or transfer risks. Along similar lines and in addition to these extensions of the risk analysis, another suggestion is to involve the experts in the real situations from different perspectives in the risk management process (from the performance of the risk analysis to the definition of risk reduction measures). The benefits of such an inclusive approach are twofold. They allow to reach beyond a theoretical analysis of the risks or definition of risk reduction measures and account for actual field experience. Besides, they provide a way to share across the organization the different views of risks and to highlight the tensions and debate when needed to come up with measures acceptable and understood by all, even if not ideal to anyone.

A second way forward consists of extending the definition of safety beyond the mere management of risks by acknowledging the inevitable contingencies and uncertainties that escape the scope of applicability of risk management. This proposal challenges the current foundation of the SMS, namely, the hierarchical risk control model. As a matter of fact, it assumes that not everything can be anticipated, addressed, and controlled a priori and acknowledges the need for adaptation to the dynamically changing situation requirements, including at the operational level. Such a perspective on safety requires a complete shift from aiming at eliminating uncertainties to trying to reduce those that can be and living with the inevitable and evolving ones. This change in philosophies challenges the current safety management frameworks. Indeed, living with uncertainties may require changing performance modes depending on the situation, making standards both internal to organizations and external, especially those used as a reference by most aviation authorities, inappropriate to safety governance. Acknowledging uncertainty calls for humbler and more inclusive and contextualized approaches.

The third proposal starts from the observation that safety is never the only stake, whether at the operational, managerial, or governance level. It is always managed at the same time as many other interests (e.g., production, liability, quality, resources). The variety of motivations of the safety stakeholders that led to the convergence toward the SMS as a new safety approach in the 1980s and 1990s provides a perfect illustration that even safety approaches fulfill many other functions than just enhancing safety. In other words, safety cannot be isolated from the rest. It exists and makes sense as part of a multidimensional and complex context. Therefore, in order to apprehend safety and manage it, it is more relevant to consider this overall context (e.g., of an organization) and look at it from a safety standpoint than to look exclusively at safety as if it were stand-alone.

Considering safety as one stake, among others, may be done in different ways. It is common to oppose safety and production, for example, or safety and finance, when safety is considered a cost. This trade-off perspective tends to put the various aspects in competition against one another in a zero-sum game. The proposed approach aims, as far as possible, to foster synergies between the various interests rather than to nurture the tensions among them. This involves adopting a more transverse perspective as opposed to individual indicators for each stake's performance. What benefits all or at least several stakes at the same time becomes a central question. This change in perspective challenges not only the current organizational structures where divisions or departments are dedicated to a given stake with their own performance objectives and criteria but also the very definition of performance. It calls for new modes of governance whereby this notion would be debated among the stakeholders, as well as the associated governance structures and practices.

7.3 A MULTIDISCIPLINARY ANALYSIS: AN ASSET TO APPREHEND COMPLEXITY

Interestingly, when I underwent in 1993 the interview to apply for the specialized master's in safety and prevention of major technological risks at Ecole Centrale Paris (a French engineering school), Alain Desroches, the former director of this education program, told me that "risk management is learning to doubt." Combining several disciplines to analyze the SMS, especially risk management and sociology, pushed me one step further in the art of doubting, by doubting the very object of risk management but constructively doing so. Indeed, it opened my reflections to other definitions of safety than that used in systems engineering based on risk management—other angles to analyze an a priori safety approach, in particular a socio-historical one. Engineering research works have focused on the SMS to try and enhance it—for example, its quantification (Li, 2019). Yet they take the SMS as a safety approach for granted and aim at improving its capabilities as such. The critical analysis of the SMS as a safety approach presented in Chapter 4, even if limited to the definition of safety as the management of risks, indeed highlights some areas of improvement in the risk analysis itself. However, calling for other definitions of safety leads to even putting into question the very premises of risk management as the main basis of safety, thus opening another world of challenges not addressed by the SMS. It is the case especially of the definition of safety proposed by organizational sociology. Safety is seen as an emerging property of organizations. This acknowledges the complexity of organizational functioning and the inevitable existence of the unexpected—of contingencies and uncertainties beyond that of probabilities.

Conversely, starting from the sociological definition of safety would not have helped address the major safety issues related to technological design that can be addressed a priori, by a lack of practical methods. Confronting the two views of safety allowed for pointing out their different scopes of validity as a basis to develop their complementarity. However, this reflection would remain within the world of safety, notwithstanding the distinct definitions. As such, combining risk management and a sociological view on safety adds to the puzzle of the adoption of the SMS in aviation as a step change in safety.

The socio-historical angle, already at the crossroads between sociology and history, allowed us to make sense of this genesis and generalization of the SMS. It did so by putting the SMS back into a wider context and highlighting its many facets beyond being a safety approach. Investigating the historical context in which the SMS appeared provides insights into its coupling with a certain technological, social, and political context that would have remained unseen otherwise. Likewise, the analysis of how ideas traveled revealed the role of communities and of some individuals, as well as the power games at play in the convergence toward what will be called a "key turning point" in safety science even though it was endorsed by safety science after the fact.

What the combination of risk management, sociology, and socio-history showed is, above all, the multiple facets of the SMS. Other perspectives, such as a legal one or a political science one, would most probably help further complete the picture. Yet the multidisciplinary approach adopted in this research already allowed for highlighting and illustrating the complexity of the emergence and spreading of the SMS. Thereby, it provided solid grounds to avoid too simple explanations or assessments of the SMS limitations and to make sense of a still commonly used approach in many high-risk industries.

7.4 THE SMS, A STEP CHANGE IN AVIATION SAFETY?

SMS, as such, is not the promising safety approach it was announced to be in aviation, where it was adopted and pushed by CAAs decades after the idea emerged in other high-risk industries. In fact, the SMS responds to many other interests beyond safety, which cannot make it an efficient safety dispositive as if it were exclusively meant to enhance safety. In reality, no approach can be since safety exists as part of a context that involves various interests interrelated with one another, but the barycenter between them may vary. In the power game that led high-risk industries to converge toward the SMS, new (at that time) scientific insights on organizational safety (or unsafety) didn't have a big share of voice.

With its practical limitations, especially the resources, both qualitative and quantitative, needed to implement it, the SMS often misses its initial intention, that of moving away from too prescriptive regulation and leaving some leeway to organizations to customize their risk analysis and define risk mitigation measures that are suited to their own context. Indeed, in organizations that are more concerned with being compliant than being safe or that consider that both are the same, developing and maintaining an SMS has become a cut-and-paste and an indicator management exercise sometimes left to an external consultant (selling similar SMSs to all its clients), with no reflexivity on the organization's specific risks, practices, and environment.

For aviation organizations that took the SMS opportunity to further push their already advanced reflection on safety (taking this opportunity reflects a certain maturity in safety), its conceptual and methodological limitations make its safety improvement potential limited as well.

Being the main approach around which safety management is regulated in aviation today, the SMS may even amplify the illusion of control—that is, the illusion that safety is actually managed as soon as an organization has an approved SMS, especially for less safety-mature organizations. Furthermore, the SMS was shown to lead to internal overregulation driven by work auditability, managerial insecurity and liability, and audit practices (Størkersen et al., 2020). Whereas it was developed to move away from sharp-end detailed prescriptions to organizational aspects of safety, it seems that it closed the loop back.

REFERENCES

Li, Y. (2019). *A systematic and quantitative approach to safety management*. https://doi.org/10.4233/uuid:458a384f-6f8a-4fc3-8bc4-c01397b54b59

Størkersen, K., Thorvaldsen, T., Kongsvik, T., & Dekker, S. (2020). How deregulation can become overregulation: An empirical study into the growth of internal bureaucracy when governments take a step back. *Safety Science*, *128*, 104772.

Epilogue

The COVID-19 Pandemic: An Amplifier Case Study

The COVID-19 pandemic that struck the whole world at the end of 2019 provides an illustration of the arguments made in this book. Of course, the situation is extreme and fortunately does not reflect everyday conditions. Yet the magnitude of the events amplifies and thus reveals phenomena that are experienced on a daily basis at a lower scale but often remain unnoticed or are not acknowledged by those who are not directly facing them. Thereby, COVID-19 offers a magnifying glass to apprehend key facets of the reality of safety management. Although a more extensive and documented research would be required, a first analysis can be performed through the lenses proposed by this book.

COVID-19: MORE THAN JUST A CRISIS

The COVID-19 episode could be considered a crisis and thereby dismissed as a case study for aviation safety management. However, the pandemic, by its magnitude, reach, and duration, has already had impacts that will be long-lasting. Furthermore, the suddenness and switch into the complete unknown phase are over. Of course, there are still a lot of things that are not understood, but the situation at the end of 2020 can no longer be considered a crisis but rather new grounds we have to live with. In one year, the world was overturned in many respects, especially more than a million deaths, restrictions on people's mobility and even lockdown in many countries, massive economic downturns partly imposed by governments forcing some activities to stop (e.g., restaurants, bars), increased social inequalities, and changes in social relationships.[1] The list could go on and tackle many other aspects.

Among the most severely hit economic activity is aviation. With the restriction of mobility enforced in many regions and countries closing their borders or imposing deterrent quarantine conditions, air traffic has dropped dramatically after decades of sustained growth.[2] How do we think and manage safety in this unprecedented context?

This section is structured along the three main proposals made in Chapter 6: (1) extending the risk analysis framework, (2) reaching beyond risk analysis as a definition of safety to encompass coping with uncertainties, and (3) addressing safety as part of a broader context.

EXTENDING THE FRAMEWORK OF RISK ANALYSIS

Although the aviation operations themselves that are still performed (e.g., flights, air traffic control) may not be so different from what they were before the COVID-19

DOI: 10.1201/9781003307167-8

pandemic, the overall context has changed dramatically. The risk analyses as they are performed in current SMS have no reason to significantly evolve. Indicators such as the number of accidents and incidents or the number of safety reviews may remain identical. Yet the COVID-19 situation leads to considering new issues that are partly novel due to the aviation context. ICAO, in its ICAO Handbook for CAAs on the Management of Aviation Safety Risks Related to COVID-19, issued in May 2020 (ICAO, 2020), identifies several topics where information should be collected to support the COVID-19-related risk analysis:

- Data on the current COVID-19 including absolute and relative rates: this includes information on the health aspects (e.g., cases, tests) but also the travel restrictions and conditions.
- Status and volume of traffic during the pandemic: this includes all the flights anticipated, including the transportation of dangerous goods or medical equipment or cargo flights performed with aircraft certified for transportation of passengers.
- Impacted operational personnel according to ICAO Annex 1—Personnel Licensing (air traffic controllers, pilots and cabin crew, aircraft maintenance engineers, flight dispatch, aeronautical meteorology personnel, etc.): this includes all the aspects related to licenses and medical certificates for crew members, or training and checking.
- Continuity of flight operations: this includes issues related to the storage and de-storage of aircraft, approvals and certificates, the availability of resources to support the activities needed to operate aircraft, and exceptional operational considerations related to crews.
- Operational status of the air navigation services (ANS) provision and limitations: this includes the availability of services and equipment, airspace limitations, or capacity reduction due to the situation.
- Aerodromes and infrastructure availability: this includes the aircraft parking positions available (even the use of other paved surfaces), location of parked aircraft, availability of critical services such as handling, fueling, customs, and so on.
- Impact on the oversight activities of the CAAs: this includes all the authority's capabilities and planning revisions during the pandemic (e.g., audits and inspections postponed), the exemptions granted in all domains, and the conditions under which differences would be acceptable to the destination states.

These categories reflect possible new hazards to operational activities of aviation stakeholders during the pandemic. As such, they are still focused on operations at a short-term horizon, considering the new operational context. As stated by ICAO, "The data collected should be used to inform the application of a risk management approach for the activities of the CAA and support the development of plans to restart operations, which will need to be a collaborative decision" (ICAO, 2020, p. 4–1).

Interestingly, in the same document, ICAO explicitly refers to ways of working: "To collect the relevant data and information, CAAs are encouraged to establish an open and continuous dialogue with all aviation domains and other stakeholders involved in tackling the pandemic, and as described in the 3 Cs [cooperate, collaborate, and communicate]" (ICAO, 2020, p. 4–1). Recognizing that the 3 Cs are vital functions in tackling global crises, ICAO adds: "CAAs should recognize that these existing functions also continuously contribute to the effective implementation of an SSP [State Safety Program], which is important in managing aviation safety risks, including the impact of COVID-19 to the aviation system" (ICAO, 2020, p. 2–1). Even though it does not tell us much about what is expected from this cooperation, collaboration, and communication or about the practicalities of these interactions, it could be an interesting entry point to address the interfaces between aviation stakeholders, provided these 3 Cs involve more than one-to-one relationships between the authorities and the other aviation organizations.

Nevertheless, it seems that the time frame for the risk analysis remains very short-term, both in the hazards and in the consequences identified. Among the material developed to support the integration into the SMS of the COVID-19-related aspects, the ACSA (Central American Aviation Safety Agency) developed three bow-tie diagrams:

- Bow-tie risk assessment on air operations impact
- Bow-tie risk assessment on the impact of airport operations
- Bow-tie risk assessment on the degradation of the capacity in the functions and responsibilities of the CAAs

A close look at the bow-tie risk assessment on air operations impact shows only two hazards considered:

- Crew/passenger infection during travel
- Transportation of infected crews and passengers by air

ICAO, also pointing to these bow-ties through the ICAO Safety Management Implementation webpage (www.unitingaviation.com/publications/safetymanagement-implementation/content/#/lessons/cHZW4YD184M_KUSyKemRIde_j6NYXjx0), includes the following note:

> These bow-ties are only examples and do not capture all "safety events" or precursors to the Top Event e.g. infection of critical flight operations personnel to support safe flight operations. Users should aim to use this as a starting point and tailor the approach for the specific operational environment.

Even with this note in mind that invites widening the scope beyond the aircraft itself to other aviation personnel, the hazards considered remain at the operational level.

As for the consequences of the COVID-19 impacts on air operations identified in the bow-tie, they include the following:

- Economic impact on the aeronautical industry
- Partial decrease in air operations worldwide
- Validity expiration of licenses, certificates, and authorizations

- Reduction in safety margins
- Public health impact
- Loss of jobs in the aeronautical sector

Critically, the economic impact on the aeronautical industry or the loss of jobs in the aeronautical sector are not, in turn, considered safety hazards, even in the long run if the aviation activity resumes. Yet the current situation in which aviation is heavily affected also leads aviation organizations to consider the future. In Europe, EASA (the European Aviation Safety Agency) envisages four areas where the industry needs to reinvent itself.[3]

- Integrate health safety as an intrinsic element of aviation safety.
- Revisit the financial framework of the industry as a whole (the examples taken are those of niche suppliers leaving the manufacturing industry with a delivery gap if they go bankrupt or of ANSPs being paid by those using the service, leaving a big financial gap when the traffic almost comes to a stop).
- Make the industry more sustainable as a response to movements such as *flygskam*, which "are starting to rally to encourage passengers to avoid flying, almost as a civic duty and commitment" (ibid.).
- Continue investing in new technologies, especially to increase "the systemic capacity of European air navigation service provision" (ibid.).

Interestingly, although the weaknesses of the aviation industry's financial model or the urge to make aviation more sustainable are identified as necessary to the survival of the aviation industry, no link is made with safety. Yet one could also envisage the impacts on the safety of massive layoffs in the aviation industry with no anticipation of knowledge transfer or competence sustainability. Likewise, the increasing concern about aviation sustainability is not considered a safety hazard. Could some decisions made with the sustainability objective in mind undermine safety, as some security decisions did after 9/11 (e.g., the cockpit door made mandatory for security reasons turned out to be a safety hazard)?

Other hazards, some on a longer-term scale, could be added to a risk analysis as illustrated hereafter.

- Internal resources available both qualitatively and quantitatively, human and technical: This results from human resources, technical, and financial decisions made today, whose impact on safety may be appreciated in the future if traffic resumes and grows beyond today's level. For example, privileging one aircraft type (and all the associated resources it implies) over the others to perform today's remaining operations may have a relative safety impact on future operations (when traffic resumes) that might differ from that of balancing today's operations between the various fleets.
- External resources available: This partly results from the economic and financial situation of the aviation industry, where some suppliers or subcontractors go bankrupt. This could affect operations (both today's and tomorrow's) both directly and indirectly, depending on the subcontracted activity.

- Uncertainty with respect to the sustainability of one's job or of the organization as a whole: How do we remain commited and focused at all levels of the organization in such conditions?
- Competence and knowledge management in a context where aviation organizations mostly lay off a part of their staff and, for some of them, mainly experienced staff: Part of the knowledge of the aviation organizations suddenly disappears to cut costs and try and survive the crisis.
- In a situation where operations are lower than they were in the cruise regime that existed a year ago and where all aviation organizations have changed their operational arrangements, new risks may arise at the interfaces.
- Maybe even more so that the regular exchanges that might have been in place prior to the crisis between these organizations may have disappeared.

The extension of the risk analysis related to the COVID-19 context could again go much further than today's operations, both in the search for hazards and in the timeframe considered for the identification of potential consequences.

Even though the ways of working on the risk management approach were touched upon in the ICAO Handbook for CAAs on the Management of Aviation Safety Risks Related to COVID-19, it does not say much about how to cooperate, collaborate and communicate nor who is concerned within organizations. It mainly addresses the interactions between CAAs and aviation organizations. Yet involving staff from all levels in the risk management process may seem even more obvious in the current COVID-19 situation. Indeed, it is an unprecedented and even unimagined situation whereby the only knowledgeable actors are those experiencing the situation. How do we approach the new hazardous situations induced by the COVID-19 situation or, for example, the risks as faced by the operational personnel by other means than involving them? For example, how does the social distancing work between pilots and other staff interacting with them directly like mechanics? Does it affect the nature and quality of the interactions and possibly of safety-critical interactions?

MANIFEST UNCERTAINTIES OR THE INCREASING EVIDENCE OF THE ILLUSION OF CONTROL

The COVID-19 episode provides a wealth of illustrations of the inevitable unknowns and unexpected events and situations. Even though the possibility of a worldwide pandemic was anticipated by some institutions like the World Health Organization, for example, all the impacts were not suspected and, even less so, their combination. As such, since it started, the COVID-19 episode represented a trip to the so-far unknowable, where a priori arrangements proved not sufficient to cope with the situation. Permanent adaptations were needed from aviation organizations (and still are) to adjust to the dynamically evolving situation at the same time sanitary, economic, political, and social. To build on Schulman et al.'s (2004) categorization of reliability professionals' modes of performance, it is the whole industry that has at all levels to cope with unprecedented system instability and option variety. Sudden decisions of countries to close their borders or impose a quarantine upon arrival or require a COVID-19 test before traveling by air, temporary financial support allocated to the aviation field under the condition to increase its sustainability, sudden decisions to close airports or reduce the air navigation service, and uncertainties on revenues and their evolution in time are illustrations of the variability that the aviation domain has been experiencing for the past year. Examples of the limitations of a priori risk management to cope with real-life contingencies have been increasing on a daily basis in aviation as in many other domains. So have the illustrations of the need to adapt one's performance mode to the context, at all levels within organizations and between organizations (see the 3 Cs principle emphasized by ICAO as a need in crisis situations but also very useful on a general basis). For example, safely transporting medical equipment in aircraft certified for the transportation of passengers required an adaptation not only of the aircraft themselves but also of the certification protocols. Likewise, lodging crews in countries imposing a quarantine required to change from the standard hotel options and yet still guarantee that crews would get sufficient and good quality rest to be able to perform return flights safely. Some aspects may have been anticipated in particular through an extension of the risk analysis beyond operations. However, the COVID-19 situation illustrates the need for combining anticipation and a priori risk management with other strategies developing flexibility and adaptability to cope with the inevitable unexpected situations and manage safety.

Furthermore, the emphasized variety of contexts calls for ad hoc measures to manage safety. Indeed, the COVID-19 situation amplifies the diversity of aviation organizations' situations, both internally and externally. Among the internal variables are, for example, the remaining resources, the financial situation, the expiration dates of licenses and certificates, and the industrial model—for example, regarding training and maintenance, either internal or subcontracted. As for the external variables affecting aviation organizations functioning and operations, they include the political decisions on citizen mobility and quarantine conditions, the availability of financial support, the social attitude toward air transport, the economic situation of potential travelers, the availability of CAAs' resources, and the situation of suppliers and subcontractors. In such a context, the need for inclusive and contextualized governance, as opposed to generic standards, becomes more obvious.

SAFETY AS PART OF A BROADER CONTEXT

As mentioned earlier, the COVID-19 episode led the EASA to identify four areas where the industry needs to reinvent itself: making health safety an intrinsic element of aviation safety, making the financial model of the industry sustainable, making aviation sustainable, and investing in new technologies.[4] Although safety, financial, and environmental considerations are on the radar to reinvent aviation, they are advanced by EASA independently from one another to respond to distinct hazards or vulnerabilities of the air transport system.

Yet the financial vulnerability highlighted by EASA—namely, the possibility of having a key supplier go bankrupt, leaving a gap in the production chain—is not disconnected from safety. Many activities directly affecting safety are totally or partially subcontracted, ranging from the production of technological safety nets or systems critical to safety to the performance of safety management activities (e.g., safety risk analysis, SMS documentation production). Along the same lines, industrial arrangements to design and produce in-house or rely, at least partially, on third parties are not neutral to safety management. Conversely, safety management may lead to suggesting certain industrial arrangements to better protect safety-critical functions and keep them flexible enough for future evolution.

The current economic situation leading to massive layoffs of aviation professionals cannot be disconnected either from today's or tomorrow's safety challenges. What competencies are needed today to ensure aviation safety in the COVID-19 context? What competencies will be needed for tomorrow, knowing that aviation involves highly trained professionals relying on long and costly training? Will aviation be able to recruit adequate resources for the future? This question also relates to the environmental reputation of aviation as further developed later.

The COVID-19 also provides an obvious illustration of the interrelation between financial and environmental aspects. Indeed, some of the financial supports provided to aviation to compensate for the dramatic reduction in traffic are subject to huge environmental efforts' condition. This condition is related to the social pressure to make aviation sustainable, as underlined by EASA, and may also apply to other sources of financing beyond governmental ones today and in the future. The social pressure for a more sustainable aviation existed before COVID-19 (e.g., the *flygskam* movement that started in Sweden in 2018), but it was even amplified by the huge traffic decrease related to the sanitary measures that "proved it possible" to live without aviation or at least without massive air traffic. This sustainability challenge directly affecting aviation's reputation has begun to show other effects beyond citizens turning their back on aviation. Indeed, young students (even in aviation universities) are starting to challenge decision-makers to push them to act urgently and massively on the sustainability front.[5] Associations of aviation students and professionals are also getting organized and increasingly active (e.g., Supaero-Decarbo). The push for sustainable aviation coming from both inside and outside the aviation industry also poses challenges in the recruitment of youngsters in the domain. Yet aviation safety requires and will continue to require competencies at all organizational levels.

As for the high priority put on aviation sustainability, in a context of scarce financial resources, where part of the money will be under scrutiny to ensure that it is dedicated to environmental efforts, managing the interrelations between environment and safety will need to be done with caution. Indeed, if both aspects are addressed as competing (e.g., for resources, in terms of social acceptability), safety could well decrease to the benefit of environmental indicators.

With the COVID-19 episode and the destabilizations it induced in many areas, the interrelations between various stakes have become more salient, as exemplified earlier.[6] Could anyone think about safety and manage it in the same way as before the COVID-19 pandemic in a context where

- the economic situation of both the aviation industry (including bankruptcies and extended low revenues) and the world economy are down (unemployment);
- resources are scarce (both financial and human, including for CAAs, even more so when their resources partly come from taxes on air traffic);

- new habits of work within and among aviation organizations and with authorities have developed with more ad hoc considerations;
- the pressure on aviation sustainability is higher than ever (social but also financial);
- the social situation of aviation (massive layoffs, uncertainty about sustainability of jobs, difficulties in recruiting) and of the world in general (remote work, social distancing) has changed;
- social expectations with respect to aviation have changed; and
- contingencies have proved a reality and total control an illusion?

Although the COVID-19 episode is extreme by the reach and magnitude of its impacts, the interrelations that it contributed to amplify existed and will continue to do. Further, in a context of scarce resources and huge uncertainties for today and the future, maintaining competition between the various organizational interests could turn out to be fatal. Very little margin is available or can be afforded. Most aviation organizations are operating on the edge in all respects. As such, the COVID-19 pandemic provides a sad and costly incentive to put safety back into context, involve safety stakeholders in the governance, and strive as much as possible to foster synergies between the various stakes. It also is a great lesson in humility.

NOTES

1. See https://coronavirus.jhu.edu/map.html.
2. According to ICAO, at the end of 2019 but prior to the COVID-19 pandemic, the traffic was anticipated to double by 2030 (www.icao.int/safety/SafetyManagement/State%20 Letters/084e%20(1).pdf).
3. See www.eurocontrol.int/article/life-beyond-covid-19-how-will-aviation-need-change? utm_campaign=coschedule&utm_source=facebook_page&utm_medium=EURO CONTROL.
4. See www.eurocontrol.int/article/life-beyond-covid-19-how-will-aviation-need-change? utm_campaign=coschedule&utm_source=facebook_page&utm_medium=EURO CONTROL.
5. See www.lemonde.fr/idees/article/2020/05/29/aeronautique-la-transition-ecologique-impose-une-profonde-transformation-de-notre-industrie_6041127_3232.html.
6. Describing the interrelations between stakes in a linear written format is a tricky exercise that can just contribute to giving a flavor of the complexity of real situations.

REFERENCES

ICAO. (2020). *Handbook for CAAs on the management of aviation safety risks related to COVID-19 issued in May 2020*. Doc. 10144. ICAO.

Schulman, P., Roe, E., Eeten, M. V., & Bruijne, M. D. (2004). High reliability and the management of critical infrastructures. *Journal of Contingencies and Crisis Management*, *12*(1), 14–28.

Appendix: Description of the Main Aviation Stakeholders

Table A.1 provides a brief description of the various stakeholders of the air transport system with an emphasis on operations and safety. The content of this table was developed over three years based on input from the ENAC "Safety Management in Aviation" advanced master's trainees. Thirty-one trainees were involved, coming from 19 different countries on 4 continents and representing a cumulated aviation experience of 167 years. It was also used as an input to the D4.1 report of the Safemode European project.

TABLE A.1
Description of the Main Aviation Stakeholders with an Emphasis on Operations and Safety

Airline	
Back-office	Organizing and managing the airline network, the flights, the personnel, and more generally, the resources (e.g., needed for line maintenance), including making sure the operational personnel is trained, licensed, and available
Real-time	Performing the flights and supporting the various ongoing flights (through the operations control center), making the necessary adjustments considering the conditions of the day
Interactions	• With air traffic management (back-office): negotiating flight schedule; (real-time): clearances, weather info • With ground handling (real-time): info on loading of the aircraft, refueling • With authorities (back-office): certification, airworthiness directives, reporting • With maintenance and repair overhaul: report malfunctions/inform the airline about actions to be taken if needed • With aircraft manufacturer: (back-office) master minimum equipment list; standard operating procedures and updates; (almost real-time) receiving technical assistance when needed[1]
ATM/ANSP (Air Traffic Management/Air Navigation Service Provider)	
Back-office	Upstream from ANSPs, the network manager ensures the strategic planning of the various flights and slots required by airlines, from a few months ahead of time to a few hours. Organizing and managing the air navigation services organization' resources (human, training devices) and their compliance with regulations
Real-time	Ensuring air traffic control of airspace under responsibility (especially clearances) Providing flight information services Ensuring communication, navigation, and surveillance Providing aeronautical information publication (AIP) and notice to airmen (NOTAMS) Weather forecast and observation Informing and assisting

(Continued)

TABLE A.1 *(Continued)*
Description of the Main Aviation Stakeholders with an Emphasis on Operations and Safety

Interactions	• With airlines: clearances, weather info • With aerodromes: traffic info, infrastructure info • With authorities (back-office): regulation and reporting • With adjacent ANSP: coordination on in-/outcoming traffic
MRO (Maintenance and Repair Overhaul)	
Back-office	Organizing and managing the MRO resources and their compliance with regulations (e.g., making sure the operational personnel is trained, licensed, and available)
Real-time	Performing the required maintenance tasks Ensuring the continuing airworthiness of an aircraft or aircraft part (including overhaul inspection, replacement, defect rectification, and compliance with AD and repairs)
Interactions	• With airlines: maintenance and engineering specifications; performing aircraft maintenance • With aircraft manufacturer: aircraft maintenance manual, instruction for modification, service bulletins, airworthiness directives • With approved training organizations (ATOs): personnel training • With authorities (back-office): regulation, licensing, reporting • With aerodrome: use of infrastructure and equipment
CAAs (Civil Aviation Authorities)	
Mission	Developing regulations, policies, guidelines, and standards Certification Surveillance
Interactions	• With ministry of transport/aviation: politics, strategic objectives • With ICAO: documentation (e.g., standards and recommended practices)[2] • With other national authorities (e.g., telecom): frequencies assignment; weather data • With investigation boards: data sharing, recommendations • With ANSP/airlines/aircraft manufacturers/MROs/ATOs/aerodromes/ground handling: regulations, guidelines, certification, approval, surveillance, personnel licensing, reporting • With other CAAs: data sharing, mutual assistance
Airport	
Back-office	Providing aeronautical infrastructure (runways, taxiways, aprons, lights, terminal) Providing radio navigation equipment (e.g., instrument landing system [ILS], localizer [LOC])
Real-time	Managing and coordinating ground services (suppliers): refueling, handling, de-icing (if need be)
Interactions	• With airlines • With passengers • With suppliers (ground handling, fuel, etc.) • With firefighters • With security authorities • With authorities (back-office): regulation, licensing, reporting, etc. • With MROs (use of infrastructure and equipment)

TABLE A.1 ***(Continued)***
Description of the Main Aviation Stakeholders with an Emphasis on Operations and Safety

ATO (Approved Training Organization)	
Back-office	Organizing and managing the training organizations' resources (human, training devices) and their compliance with regulations (e.g., making sure the instructors are trained and licensed and full-flight simulators or other simulation devices are certified) Develop training programs compliant with the aircraft training manual
Real-time	Deliver training and exams
Interactions	• Aircraft manufacturer: aircraft flying training manual • Simulators suppliers • CAAs: approval of ATO, reporting • Airlines • MROs
OEM (Original Equipment Manufacturer; e.g., Aircraft or Engine Manufacturer)	
Back-office	Design and production of aircraft, supplemental type certificates (STCs), upgrades[3] Development of aircraft technical documentation (MMEL, maintenance program, flight crew operating manual) Certification of aircraft, technical documentation, STC approval, upgrades
Real-time	Support to airlines in case of aircraft on ground (AOG), engineering question
Interactions	• Airlines: including the design of aircraft • CAAs: certification, reporting of significant events • MROs: maintenance program and procedures • ATOs: aircraft training program and procedures
Ground Services	
Back-office	Organizing and managing the ground services organizations' resources (human, tooling, equipment) and their compliance with regulation (e.g., making sure the operational personnel is trained, licensed, and available)
Real-time	Ensuring handling services: cabin servicing, handling, ramp services, check-in counter services Ensuring fueling and de-icing Ensuring security services for handling activities: catering escort, setting security perimeter around the aircraft
Interactions	• Airlines • Airport • Passengers • Fuel suppliers • CAAs • Immigration services • Police • Customs

NOTES

1. This document, developed by the aircraft manufacturer, specifies the list of equipment that at a minimum need to be operational for the aircraft to perform a flight in safe conditions.
2. International Civil Aviation Organization.
3. Authority approved major modification or repair to an aircraft.

Index

Note: Page numbers in *italics* indicates figures and page numbers in **bold** indicates tables.

Printed in the United States
by Baker & Taylor Publisher Services

Printed in the United States
by Baker & Taylor Publisher Services